WHY BUILDINGS STAND UP

The Strength of Architecture

WHY BUILDINGS STAND UP

The Strength of Architecture

MARIO SALVADORI

Illustrations by
SARALINDA HOOKER
and CHRISTOPHER RAGUS

McGRAW-HILL BOOK COMPANY

New York St. Louis San Francisco Bogotá Guatemala Hamburg
Lisbon Madrid Mexico Montreal Panama Paris San Juan
São Paulo Tokyo Toronto

Reprinted by arrangement with W. W. Norton & Company, Inc.

First McGraw-Hill Paperback Edition, 1982

567890 RRD RRD 87654

ISBN 0-07-054482-4

Designed by Jacques Chazaud
Composition by Spartan Typographers

Library of Congress Cataloging in Publication Data

Salvadori, Mario George, 1907–
Why buildings stand up.

Reprint. Originally published: 1st ed. New York
New York : Norton, 1980.
Includes index.
1. Structural engineering. I. Hooker, Saralinda.
II. Ragus, Christopher. III. Title.
TH845.S33 1982 624.1′7 81-15645
ISBN 0-07-054482-4 (pbk.) AACR2

To Carol,
my wife

Contents

Preface

This book was written for those who love beautiful buildings and wonder how they stand up.

Inspired by the constructive instinct of man, I have studied, taught, and designed structures for a lifetime. I now write about them in the hope of sharing with others my excitement and their beauty.

This book offers the reader both the history of some of the great monuments of architecture and an explanation of why they stand up. Chapters on architectural technology usually precede those about the monuments made possible by a particular technology, but some may prefer to read about architectural masterpieces before finding out why they stand up.

Since great architecture is the successful marriage of art and technology, the enjoyment of the story I have to tell does not depend on how the book is read.

Mario Salvadori

ACKNOWLEDGMENTS

I wish to express my gratitude to:

George R. Collins, Professor of Art History at Columbia University, for reading the entire manuscript, and giving me invaluable information and bibliographical sources for the historical chapters;

Stephen Gardner, my colleague at Columbia University, without whose scholarly help I could not have written the chapter on the Cathedral at Beauvais;

Stephen Murray of Indiana University, for allowing me to read his study of the reconstruction of the Beauvais Cathedral before it was published;

Saralinda Hooker and Christopher Ragus for their illustrations, which have given clarity and significance to the text;

David Christie of Hardesty and Hanover in New York, for the use of the bridge photographs used in Figures 9.15 and 9.16;

The M.I.T. Press for permission to use Figures 7.17 and 16.6 from *Developments in Structural Form* by Roland J. Mainstone, as an inspiration for the drawings of Figures 13.10 and 14.4 in this book;

John Wiley & Sons for permission to use Figures 3.29 and 5.1 from *The Masterbuilders* by Henry J. Cowan as an inspiration for the drawings of Figures 13.5 and 14.4 in this book;

The American Society of Civil Engineers for permission to reprint in Chapter 17 some material from their Preprint No. 2823, ASCE Convention, Dallas, Texas, April 25–29, 1977;

Shirley Branner, for permission to use the drawing by the late Professor Robert Branner of the reconstruction of the interior of the Beauvais Cathedral;

Kenneth Snelson, for permission to use a photograph of one of his sculptures;

Edwin F. Barber, my editor at W.W. Norton, for his patience and dedication;

Frederick Noonan, for his particular and intelligent reading of the manuscript; and

Pearl Kaufman, for her assistance in putting the book together with interest and skill.

WHY BUILDINGS STAND UP

The Strength of Architecture

1 Structures

The Beginning of Architecture

Compared to other human activities, architecture is a young art that had its beginnings only 10,000 years ago when men and women, having discovered agriculture and husbandry, were able to give up roaming the surface of the earth in search of food. Until then they had been exposed to the weather, precariously protected by tents of animal skins. Perpetually on the move, they cooked over campfires and gathered in small tribes.

All of this changed when people became sedentary. Tents were supplanted by more substantial abodes, and a permanent hearth became the center of the home. Numerous huts sprang up in fertile areas; contact between families became more frequent and intimate; villages grew. From village to village a network of paths was worn. At times paths had to cross rivers and ravines, requiring the construction of footbridges made out of tree trunks or suspended from ropes of vegetable fibers.

The clustering of huts created the need for larger huts where village problems could be discussed. These larger structures served both as town halls and churches, since spiritual needs have always gone hand in hand with the physical. Indeed the larger monuments of archeological architecture were often motivated by spiritual needs.

The last ten thousand years spanned more than 300 generations, but we who have witnessed the incredible changes brought to our cultures by the industrial revolution may feel that architecture has not changed much, at least over the last 6,000 years. This constancy in the built environment should not surprise us, if we realize that architecture satisfies basic physiological needs, which have not changed since Homo sapiens appeared about 3 million years ago. We eat the same kinds of food as our prehistoric ancestors, and we cook food much as they did. We sleep on horizontal surfaces (though surely softer today than then), we protect ourselves from the weather, and we procreate in the only way we can. Architecture is the most conservative of the human arts and sciences because it caters to these unchanging needs of man. Even our spiritual needs, which may have changed somewhat and may have produced different rituals, are as basic today as they were in prehistory.

Changes in architecture, more quantitative than qualitative, have been motivated by the conglomeration of people. The city is a friend to architecture. Whether we gathered first in villages, and then towns and cities, the better to defend ourselves from enemies, or whether the exchange of trades and crafts required the proximity of first hundreds, and eventually millions, it was the city that led us to erect taller and taller buildings and to enclose larger and larger spaces. In 2000 B.C. Minoan cities on Crete already boasted four-story houses, and tenements in the most popular sections of Rome had risen as high as ten stories.

We are the heirs of these builders. We think of ourselves as the most individualistic human beings in history and yet we gather in large halls to see the same spectacles and live in beehives containing hundreds of identical apartments. Nostalgic as we may be for the simple life of the forest, most of us find the ways of the city more congenial and more efficient. In the United States over seventy-five percent of the population live in cities of more than 100,000 people; there are at present in the world twenty cities with over seven million people, of which three (New York, Chicago, and Los Angeles) are in the United States. We have become members of group cultures. The relative isolation of the countryside is no longer our way of life.

Science and technology at their best are motivated to satisfy genuine human needs. If architecture has never changed much in its functional aspects, it has undergone a fantastic technical revolution. The needs of the city will be satisfied, and technology, spurred by the discoveries and inventions of the industrial and scientific revolutions, has come to help.

Function and Structure

The purpose of a building is to perform a function. The function of most buildings is to protect people from the weather by creating enclosed but interconnected spaces. These spaces may be many and small, as in apartment houses, or few, perhaps even a single space, as in a church or theatre. The *function* of the building is fulfilled by the construction of surfaces, like walls and roofs, which separate the outside from the inside. But walls must be pierced by doors to let people in and out and by windows to allow the penetration of light and air. Roofs must prevent rain and snow from entering buildings. By analogy with the human body the *functional* envelope of a building is called its *skin*. Within this skin the separation of internal spaces demands the construction of floors and partitions while circulation between floors requires stairs and elevators.

The *structural components* of a building assure that the elements required to fulfill its function will stand up. Columns, beams, and floors—structure—make possible the architectural function. Even in the tents of our ancestors, the functional and the structural components were clearly separated. Animal skins created the inner space and protected the dwellers; they were the envelope or functional component. The center pole and the ropes staying it made sure that the skins would stand up; they were the structural components. Again, by analogy with the human body, these structural components are called the *skeleton* or *frame* of the building.

It is in the development of structure that architecture has undergone a revolution. Our high-rise buildings now reaching heights of nearly 1,500 feet and our covered stadiums sometimes spanning 700 feet require structures incomparably more complex and stronger than those of the past. Their development demanded both new theoretical knowledge for design and new materials for construction.

In the past structures were erected by the time-honored method of trial and error. Each builder would dare a little more than his predecessor until, faced by failure, he knew that the limits of his structural system had been exceeded. The daring brick dome of Hagia Sophia in Constantinople, first built in A.D. 537 and one of the largest domes of antiquity, fell twice before it finally stood. It stands to this day. The exceptionally high vault of the Gothic cathedral at Beauvais collapsed twice before the master masons of the fourteenth and sixteenth centuries understood the real potential and limitations of this type of construction. Although Leonardo da Vinci at the end of the fifteenth century knew exactly how a beam works, the great physicist Galileo, often called the father of modern physics, had a false understanding of beams. He thought, for example, that a *cantilevered beam* (such as those supporting balconies) has a tendency under the action of the loads on it to rotate around the edge of

1.1 **Galileo's Beam Theory (from** Due Nuove Scienze, **1658)**

its lower surface (B in Fig. 1.1). Leonardo da Vinci knew 150 years before him that such a beam, instead, tends to rotate around an axis halfway between its lower and upper surface.*

Today the mathematical theory of structures is an essential part of physics and has made possible the record-breaking structures of our era. These triumphs in structural design have been achieved through the use of computers, electronic marvels with the capacity of performing millions of operations per second, while simultaneously making logical decisions.** The designs thus made possible were unobtainable only twenty years ago, not for lack of theoretical knowledge, but because they would have required years or even centuries of hand calculations. Yet the "infallible" computer is run by fallible human beings and cannot be trusted to give right answers all the time. No structural engineer accepts the output of a computer run unless it agrees (more or less) with what experience tells him to be the correct answer.

The development of structural material has not kept pace with the needs for the realization of advanced theoretical concepts. Except for reinforced and prestressed concrete and high-strength steel, the materials we use today are very similar to those used by our forefathers. Wood, stone, masonry and bricks still dominate construction and must be used in ways compatible with their properties, which have remained practically unchanged for centuries. Thus, while mankind's aspiration to reach the sky, the "Tower of Babel Complex," drives us to erect higher and higher buildings, our tallest (the Sears Tower in Chicago) is only three times taller than the 5,000-year-old pyramid of Cheops. Our largest hall (the Louisiana Superdome) spans a distance of 680 feet, which is only four-and-a-half times the 142-foot span of the dome of the Pantheon built in Rome 1,800 years ago.

The superiority of modern materials however is of a striking nature, economically if not dimensionally. Our large buildings are extremely light and are cheaper to build than those of the past. The dome of St. Peter's, the largest church of Christendom, spans 137 feet, consists of not one but two domes of brick, and weighs about 450 pounds per square foot. The dome of the C.N.I.T.*** exhibition hall in Paris, a double dome made of

* This was only rediscovered at the beginning of the nineteenth century by the French mathematical physicist Navier.

** The computer can, for instance, compare the result of a calculation with a given number and, depending on whether the result is smaller than, equal to, or greater than the given number, will continue the calculation, stop, or repeat the calculation, incorporating in it the obtained result.

*** Centre National des Industries et des Techniques.

concrete and five times larger than St. Peter's, weighs only 90 pounds per square foot.

One might suppose, then, that the shape and size of buildings would be determined essentially by the availability of structural materials. Though this is mostly true in modern times, it is fascinating to realize that religious and spiritual tenets have had an enormous influence on traditional construction. For example, the Patagonian Indians, who lived at the tip of South America in one of the worst climates in the world, exposed to a permanent and infernal wind blowing from the South Pole, were capable of building large stone domes for their churches. Constructing a large dome of heavy stone blocks without the help of mechanical devices is no mean structural achievement and one would think that the Patagonians could have used this knowledge to protect themselves from their hostile environment. But such was their respect for the gods that they would not dream of building their houses in the shape of churches. While their gods rested warmly in the stone-domed churches, the Patagonians lived in the open air and slept behind windbreakers made of vertical stone slabs. Similarly, in Arab countries many traditional houses, which should have been sited according to exposure to the sun, were often oriented by adherence to religious laws such as those establishing that the praying niche or *mihrab* in a mosque must face Mecca.

Since in most buildings functional and structural purposes are achieved through different components, the components are usually made of different materials. The so-called *curtain walls* of our high-rise buildings consist of thin, vertical metal struts or *mullions*, which encase the large glass panels constituting most of the wall surface. The curtain wall, built for lighting and temperature-conditioning purposes, does not have the strength to stand by itself and is supported by a frame of steel or concrete, which constitutes the structure of the building. On the other hand, both in traditional architecture and in modern designs, particularly of exceptionally large halls, the structure may become so all-important as to take over the functional demands of the building as well. Such is the case in large spaces covered by concrete domes. In traditional brick construction the walls have window openings and doors and act as both structural and functional components. Typical of the integration of structure and function is the traditional housing in certain towns of southern Italy, where the *trulli* are round, domed, and stone-built structures which do not have separate functional components (Fig. 1.2).

Modern construction has obtained its greatest triumphs where load carrying is not hampered by functional requirements. Bridges, for ex-

1.2 THE TRULLI OF SOUTHERN ITALY

ample, have the sole purpose of connecting two otherwise separated points in order to permit traffic between them. During the Roman Empire, all roads led to Rome because the Romans linked the farthest provinces to the capital through 50,000 miles of them. Travel from London and Baghdad to the Eternal City proceeded by crossing innumerable rivers on a type of bridge the Romans made popular (although they did not invent it)—the brick or stone arch bridge, capable of spanning up to 100 feet. The manufacture of steel cables, with the highest strength ever achieved by a structural material, has allowed the modern engineer to design suspension bridges which span almost 6,000 feet. We are doing sixty times better than the Romans in bridge design, while we are limited to relatively minor dimensional improvements in other types of construction.

Is there hope to go much beyond these recent achievements? If we take into consideration economic factors and limit ourselves to steel, the only way to go beyond our present bounds is to improve the strength of steel. This is feasible, but only within the limits of steel molecular attraction. In building, as in all of life, human endeavors are thwarted by the laws of nature.

Architects and Engineers

Even though the functional and structural components of architecture are most often distinct, structure has always had a decisive influence on architecture. In the first place, it is unavoidable. Secondly, structure must obey the laws of nature and cannot always accommodate the desire of the architect. Thirdly, structure, while necessary, is often hidden and does not appear to contribute to the architecture it supports: it is to the architect what the lawyer is to the accused, a necessary evil. Finally, structure is costly. Not really, in comparison with the cost of the rest of the building, but costly all the same. In most buildings the cost of the structure is one-fourth to one-fifth of the total cost, but in some, as in a bridge or a very large hall, it is the main cost component.

Structure is often a cause of friction in the relationship between the architect and his structural engineer. A good architect today must be a generalist, well versed in space distribution, construction techniques, and electrical and mechanical systems, but also knowledgeable in financing, real estate, human behavior, and social conduct. In addition, he is an artist, entitled to the expression of his aesthetic tenets. He must know about so many specialties that he is sometimes said to know nothing about everything. The engineer, on the other hand, is by training and mental make-up a pragmatist. He is an expert in certain specific aspects of engineering and in those aspects only. There are today not only structural engineers, but structural engineers who specialize only in concrete design or only in the design of concrete domes or even in the design of concrete domes of one particular shape. No wonder the engineer is said to be a man who knows everything about nothing! The personalities of these two are bound to clash. Lucky is the client whose architect understands structure and whose structural engineer appreciates the aesthetics of architecture. In the last resort, the architect is the leader of the construction team and to him accrue the responsibility and the glory of the project. The engineers are his servants. One of the main reasons for their employment is that, as the saying goes, an engineer is a fool who can do for one buck what any other fool can do for two.

Structure and the Layman

Where does the layman stand in the middle of this professional controversy? He is the man for whom the building is being built or one of the thousands of people who will use the building. He should be the most important, even if a seldom-heard member of the team. Luckily the interest of people in architecture has increased dramatically during the last few decades. We have learned from the psychologist that our environment influences greatly the efficiency of our work and the enjoyment of our lives. The sociologist has taught us that the design of our buildings induces or prevents crime. The environmentalist has proved that construction creates pollution. The economist has emphasized how the energy needed to condition a building depends substantially on its design. The architectural historian has told us that the image of our city is the image of our culture. The layman is not only conscious of the external appearance and of the aesthetic impact of buildings but realizes the difference between a well-designed and a poorly designed apartment. He is concerned with minimizing pollution and saving energy. Today all the essential aspects of architecture involve the layman and he participates, with increasing interest, in the development of our cities, buildings, parks, and monuments.

On the other hand, most laymen seldom look at architectural structures, or ask the simple question, "What makes buildings stand up?" This apparent lack of interest in structures is due to a misapprehension: that an understanding of structure requires a scientific mind and the acquisition of technical knowledge usually outside the province of ordinary citizens. This fear is unjustified. Structures, even large and daring structures, were built in the past by craftsmen who had no theoretical knowledge and moved on a purely physical intuition of structural principles. We all possess these intuitions through our daily experience. We understand why columns at the bottom of a building must be larger than those at its top, since they must support the accumulated weights of all the floors of the building. We understand how an arch works as Leonardo described it: "An arch consists of two weaknesses which leaning one against the other make a strength." Without any theoretical knowledge, we are ready to say that a cantilever beam, like that supporting a balcony, is "right" if shaped with a decreasing depth towards its tip and "wrong" if its supported end is less deep than its tip (Fig. 1.3). We may even have aesthetic feelings about this matter and say that the beam is in the first case "lovely" and "ugly" in the second. Moreover, we know that the physical laws of structure have not changed over the centuries and that the 5,840-feet-long Akashi-Kaikyo suspension bridge, being built in Japan, works exactly as the vegetable-fiber footbridges built in Central America 1,000 years ago.

Another subtle reason attracts us to the world of structures. A structure is an artifact expressing one of the many aspects of human creativity, but it is an artifact that cannot be created without a deep respect for the laws of nature. A beautiful structure is the concrete revelation of nature's laws. When the famous Italian engineer Pier Luigi Nervi, one of the pioneers of modern concrete and the last of the great

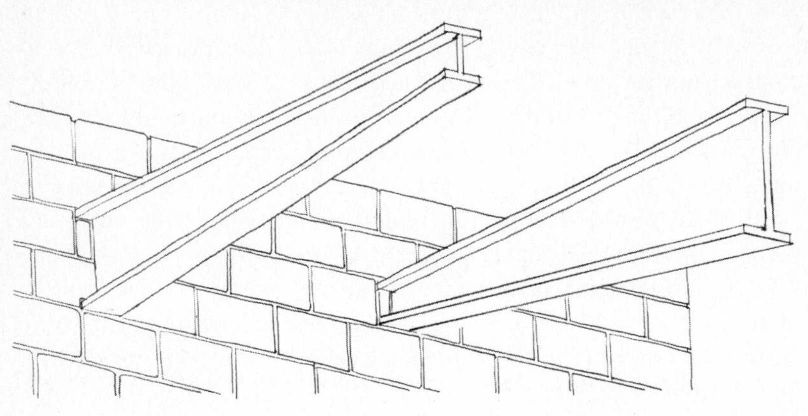

1.3 CORRECT AND INCORRECT CANTILEVER BEAMS

master builders, saw the George Washington Bridge in New York, his first words were: "To think that before it was built man had never seen such lovely curves!" This, of course, does not imply that a correct structure is necessarily beautiful, but that structural beauty cannot exist without structural correctness.

Architecture, besides fulfilling a function, sends a message to the onlooker through its varied and significant forms. No passerby confuses a church with a jail. It is perhaps not so obvious that structure too has a message of its own: it can be a message of strength or elegance, of waste or economy, of ugliness or beauty. But without it, architecture cannot exist.

Thus, in addition to speaking to us about usefulness, economics, energy, and safety, structure asks us to appreciate creativity and beauty. It is no wonder that some of the greatest minds of the past have given themselves to the study and the creation of structure and that all of us, more or less consciously, are interested in discovering the mysteries of its laws.

2 The Pyramids

What Are They?

They seem to have always been there, on the West Bank over the Nile, these man-made mountains. But what are they? Tombs, certainly. Perhaps also cenotaphs, that is, monuments to the Pharaohs which were not supposed to contain their bodies. Or astronomical instruments establishing the true North. Or gigantic public works destined to give employment to hundreds of thousands of peasants during the autumn floods of the Nile, when the strip of land along a thousand miles of its course is fertilized, but cannot be cultivated. Or temples to the Sun-god, worshipped by the powerful elite of the Heliopolis priesthood, which had defeated the priesthood at Memphis and ruled the country through the figure-head of the Pharaoh. Most probably they were or became all this and more, since they were constructed over a period of 2,500 years, but mainly in a mere 500 years during the Old Kingdom from 2,686 B.C. to 2,181 B.C.

Three pyramids reign supreme over the eighty known to exist today: the pyramid of Cheops, as Herodotus called him, or Khufu, to call him by his Egyptian name, the pyramid of Chephren (Khafre) only twenty-two feet shorter than Cheops's, but fifteen percent smaller in bulk, and that of Mycerinus (Menkaure), only a tenth as large as the other two. These grand structures constitute the greatest group of monuments built by man and the most celebrated in history. Religious monuments of great awe to the Egyptians, they sit silent in the silence of the African desert—

thirteen million tons of limestone blocks, mystery of mysteries, sending out a message perpetually changing over the centuries, the oldest monuments in the world and at the same time the largest. Fortresses unassailable to the attacks of the Moslems in search of their mythical treasures. Inspiration of glory to the troops of Napoleon, who incited his soldiers by saying: "From their tops three thousand years of history look at you." Coveted goal of adventurers and archeologists in search of the entrance to the King's chambers. Bank vaults of incredible loot to the successful thieves of the Middle Kingdom, who from 2133 B.C. to 1786 B.C. penetrated the well-hidden chambers and stole thousands upon thousands of golden objects, compelling the Heliopolis priests to plug again and again their entrances against these undaunted, sacreligious seekers of worldly goods. Source of magic information to this day and of deathly stories concerning whoever dares to invade their mystic kingdom.

Cultures are immortalized by monuments, which express their conception of the world, of life and death. It is no exaggeration to say that the Egyptians were obsessed with death. Their theology dealt with hundreds of gods, each one of which supervised this or that life activity, this or that natural phenomenon, but none was as supreme as the god that governed the afterlife. This god did not always remain the same, since his importance depended on the political fortunes of the elite ruling the kingdom. For centuries he was represented by the *benben,* a conical stone object, and called Re. But under the name of Horus, the falcon, and of Khepri, the scarab, he carried the sun across the sky, Horus flying on his immense wings and Khepri pushing the sun with his front legs, while Re himself sailed through the sky in a boat. The people and the priests of Egypt found no contradition in this triple conception of the same god. What mattered was that the Pharaoh, who united in a single kingdom the Upper and the Lower Nile, upon his death became identified with the Sun-god and, in order to live happily in the afterlife, had to be supplied, royally supplied, with all the implements of daily life. This prerogative, at first limited to the King, was slowly extended to his immediate family, then to his ministers and priests, and finally to all of his people. Thus, the concern for his happy survival became the democratic concern of every Egyptian.

The body of the King was mummified. His internal organs, except for his heart, were enclosed in separate jars, often located in a separate chamber, and his skin and bones were soaked at length in a solution of bicarbonate of soda. Completely wrapped with linen bandages impregnated with a resin or gum, his remains were enclosed in a coffin with the shape of a human body and set in a sarcophagus of stone. The mummy, together with the implements of daily usage and animal mummies, had to be protected from outside interference. Originally, the sarcophagus was set in an enclosed room, covered by an earth mound, called a *mastaba*. But through the centuries, the mastaba became a large rectangular mound surmounted by smaller mounds, until the tomb took the appearance of a stepped pyramid, similar to the ziggurats of Babylon and Assyria. As we approach the time of the IV Dynasty, the stepped pyramid disappears and the purely geometric pyramid, with a square base and four triangular sides, takes its place. It is said that at sunset, the sun rays over the Nile often pierce the clouds of a heavy sky and project a pyramid of light onto the land. The pyramid, according to one interpretation, is the perpetuation in stone of this light shape and symbolizes the aspiration of man to move towards the Sun-god.

Whether the well-preserved body of the King did actually live after death or whether his *ka* (a word whose nearest approximation is "soul") did wander in the afterlife and needed the body's preservation for its existence, has not been ascertained. But whatever the theological or cosmogonical reasons for the elaborate burial of their King, the fact remains that the most massive monuments the world has ever known were erected by a civilization without a metal stronger than copper and with no wheels or tackles, using a considerable part of its total economy and energy in this activity. Setting aside the mystery of human behavior, we cannot avoid being obsessed by a simpler question: "How did they do it?"

The Building of the Pyramids

In magnitude of construction no pyramid is more amazing than the Great Pyramid at Gizeh, erected by and for Cheops, the largest pyramid of them all (Fig. 2.1). A few numbers will make this clear. The Great Pyramid was built around a center mound of local stone and consists of a core of hewn local blocks and an outer facing of Tura limestone, containing an estimated 2,300,000 blocks of an average weight of two-and-a-half tons each, a total of 6.5 million tons of stone. Some of these blocks weigh as much as 15 to 20 tons. Its volume could easily contain the Cathedral of Milan, the Church of Santa Maria del Fiore in Florence, and Saint Peter's in Rome (the largest church of Christendom), plus Westminster Abbey and Saint Paul's Cathedral in London. Napoleon himself, comforted by the opinion of the great mathematician Monge, computed that the three pyramids at Gizeh had sufficient stone to build a

2.1 THE PYRAMIDS AT GIZEH

wall ten feet high and one foot wide around the entire boundary of France. The Great Pyramid has a square base with sides of 756 feet and covers over thirteen acres. Its original height was 481 feet. Hence, it is two thirds as high as the first American high-rise (the Woolworth Building in New York), but would occupy ten New York City blocks, rather than one.

Perhaps more amazing than its size is the accuracy with which the pyramid was built. The lengths of the four sides of its base differ at most by eight inches or, to put it differently, their lengths have an error of 1 in 1,134. These sides are oriented to the four cardinal points with a maximum error of five-and-one-half minutes of a degree or of about 1 in 4,000. The four right angles of its base are off by not more than three-and-a-half minutes of a degree, with an error of less than 1 in 1,500.

The faces of the Great Pyramid are inclined at an angle of 52° to the horizontal, an angle that is found in all other pyramids except two, in which the faces are sloped at 43 ½°. A lot has been made of these two angles, because they seem to imply that to measure them so accurately, the Egyptians had to know the value of the ratio of the diameter to the circumference of the circle, the famous Greek number π (pi), with an accuracy of 1 in 1000. (It is known, instead, that they estimated it to be equal to 3, the same value given in the Bible, with an error of about five percent.*) That this was not necessarily so has been suggested by the electronics engineer T.E. Collins, who noticed that if the Egyptians had measured the sides of the pyramid's base by a rolling drum, the pyramid side would be measured by the drum's *circumference* times the number of its revolutions. The 52° slope of its faces would then result from a pyramid height equal to 4 times the drum's diameter times the

* The Greeks' π had an error of less than 1 in 10,000,000.

number of revolutions (Fig. 2.2). Thus the pyramid half-side would necessarily equal the height times π over 4 without any need to know the exact value of π. This might sound like pure coincidence, were it not for the fact that the only other slope angle of 43 ½° results from a height of 3 times the number of revolutions times the drum diameter, and a half-side necessarily equal to the height times, again, π over 3.* The Egyptians were not very sophisticated mathematicians and it is difficult to believe that they might even have conceived of the number π. Mr. Collins's explanation may destroy one more of the many Egyptian myths, but is quite convincing.

The pyramid blocks were of three kinds: the interior blocks of local limestone, recognizable by their reddish color; the casing blocks to be used on its surface, which were cut with great accuracy out of the same stone; and the facing blocks of Tura white limestone, quarried up the Nile and carried to the immediate vicinity of the site by boat. These last blocks were smoothed to a fine finish by gangs of skilled masons, who

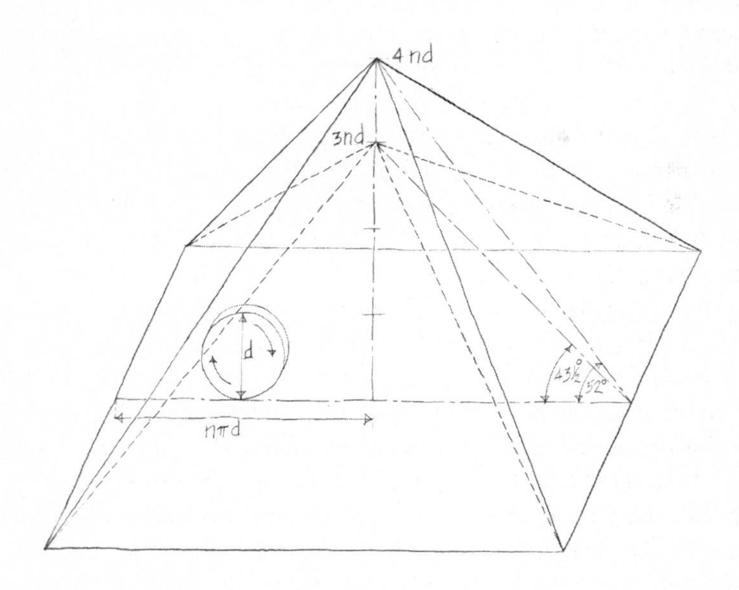

2.2 THE PYRAMIDS AND π

* In terms of trigonometry, in the first case, calling n the number of drum revolutions and d its diameter, $\tan \alpha = 4nd/\pi nd = 4/\pi$, from which $\alpha = 51°.85$. In the second, $\tan \alpha = 3nd/\pi nd = 3/\pi$, from which $\alpha = 43°.68$.

left their names on the blocks: "Stepped Pyramid Gang," "Boat Gang," "Vigorous Gang," "Enduring Gang." Their surfaces were so perfectly planed that a postcard cannot be inserted between them. The cutting of stones at the quarries was done either with copper tools (which might have been made harder by an unknown process) or, more probably, by repeated pounding with a ball of diorite (a very hard igneous rock) on the softer limestone and by inserting wedges of wood in the cuts thus produced. The wood, wetted and swelled, cracked the limestone along vertical and horizontal planes, as shown by the old quarries still in existence (Fig. 2.3). Granite was too hard to be cut by any of these methods. It was heated along straight lines by fires and then suddenly cooled with water so as to produce plane cracks by thermal shrinking. Granite slabs as large as 30 feet by 17 feet were used in the King's chamber of the pyramid of Cheops, that has a capstone also of granite blocks.

2.3 THE PYRAMIDS, STONE QUARRIES

Wheels were not used to transport the blocks to the site; the blocks were set on sledges and pulled by men over land. This was no mean feat: while the average block weighed 2.5 tons, the granite slabs in the King's chamber of the Cheops Pyramid weigh 50 tons and some of the blocks in the Mortuary Temple of Mycerinus at the foot of the ramp leading from the Nile to the pyramid weigh 200 tons. Navigation by boat from Aswan was made dangerous by the Nile's rapids, but the long voyage by water required a minor effort in comparison with the short trip over land.

While the materials were quarried and brought to the site, the site was levelled to an unusual degree of accuracy. The greatest difference in level between the four corners at the base of the Great Pyramid amounts to only half an inch, an error of 1 in 18,000. This miraculous accuracy was

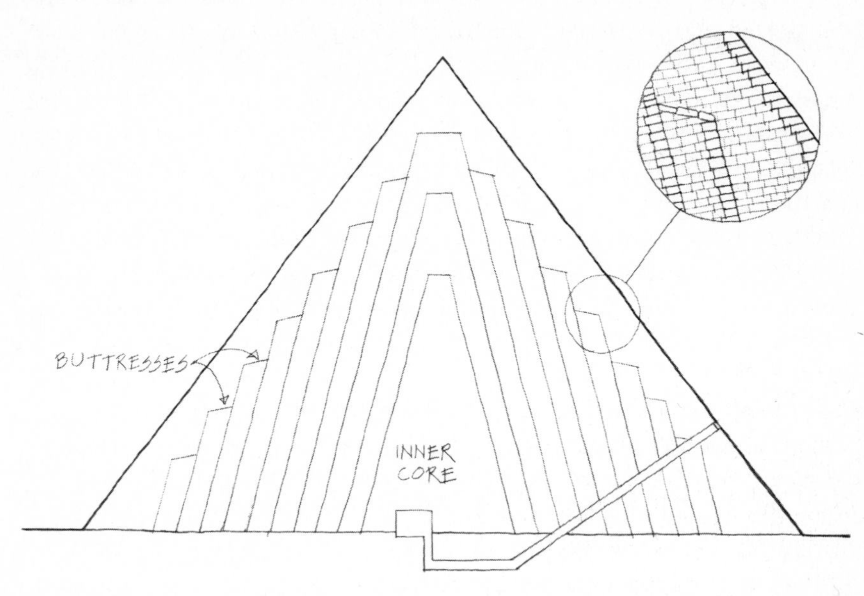

2.4 CROSS-SECTION OF PYRAMID

obtained by leveling roughly the terrain to a level lower than the surrounding area, excavating in it a network of channels and filling the foundations with water. By measuring depth from the horizontal water level and excavating all the channels to the same depth, an accurate bottom level was established for the channels, from which the remaining area of the foundation could be levelled.

While the design of some of the early pyramids was altered during construction and some of them were enlarged in successive phases, the largest were built by a fairly uniform procedure, mythically attributed to Imhotep, the great engineer of the pyramids, whose name was immortalized in the hieroglyphics of many a stone wall. (Imhotep was a physicist, a mathematician, an engineer and the inventor of building in stone. He eventually became a god and was venerated for over 3,000 years in Egyptian history.) The inner core of the pyramid (Fig. 2.4) was surrounded by masonry buttress walls, inclined inward at an angle of 75° with decreasing heights towards the exterior of the pyramid. The decreasing heights of the buttress walls gave the pyramid, during construction, a stepped shape, which is also the final shape in one of the oldest pyramids, the Stepped Pyramid of Zoser. The outer casing stones were set in layers not perfectly horizontal, but slightly inclined inward to increase the stability of this outer layer. Finally, the facing stones, with their outer surface cut at a slope of 52° (or 43 ½°) were set, also slightly inclined inward and, in the case of the Great Pyramid, in courses slightly curved downwards, that is, lower at the center of the face than at the edges, again to reduce the danger of sliding of these blocks. It is not known whether a ramp of access for the blocks was built for each face or whether, as the pyramid grew, a single feeding ramp was built around the pyramid in the shape of a square helix. There is no doubt that the blocks were pulled up the ramps by man power, as shown by the bas-reliefs on the walls of the surrounding temples. The facing blocks were set starting with the granite cap at the top and coming down the sides of the pyramid, while dismantling the ramps of earth and bricks.

The sole official purpose of the pyramid was to hide the King's chambers. These were set either in the rock under the pyramid or in the body of the pyramid itself and were reached by narrow passages dug through after the construction of the pyramid. Some of these passages are inclined downward to reach the chambers set below the pyramid, others are inclined upward to reach the chambers in its body. In the Great Pyramid, the passage leading to the King's chamber becomes the Grand Gallery, 153 feet long and 28 feet high, covered by a corbelled roof* of slabs each extending 3 inches inward from the slab below it. The King's chamber is entirely covered with granite slabs and measures 34 feet by 17 feet. It is 19 feet high. Its roof consists of nine slabs weighing 400 tons, over which are built five smaller chambers, the last of which has a roof of inclined slabs. One may question how such a stone roof can support the weight of 400 feet of stone above it, but one must realize that the superimposed enormous weight does not flow vertically down the pyramid. Just as in a river the flow of water is diverted by the piers of a bridge, the chamber cavities in the body of the pyramid divert the flow of compression due to the stones' weight producing an arch effect that lightens the load to be supported by the chamber's roof (Fig. 2.5). It is still true that the pressure on the pyramid's stones increases with its height: the volume and hence the weight of the stones increases as the cube of the height, while the area of the surface on which it is supported

* A corbelled roof is a fake arch obtained by cantilevering the stones of each layer inward from those of the layer below it. The cantilevered layers are kept in place by the weight of the layers above them. Corbelled stone arches and domes appear in some of the oldest monuments built thousands of years B.C.

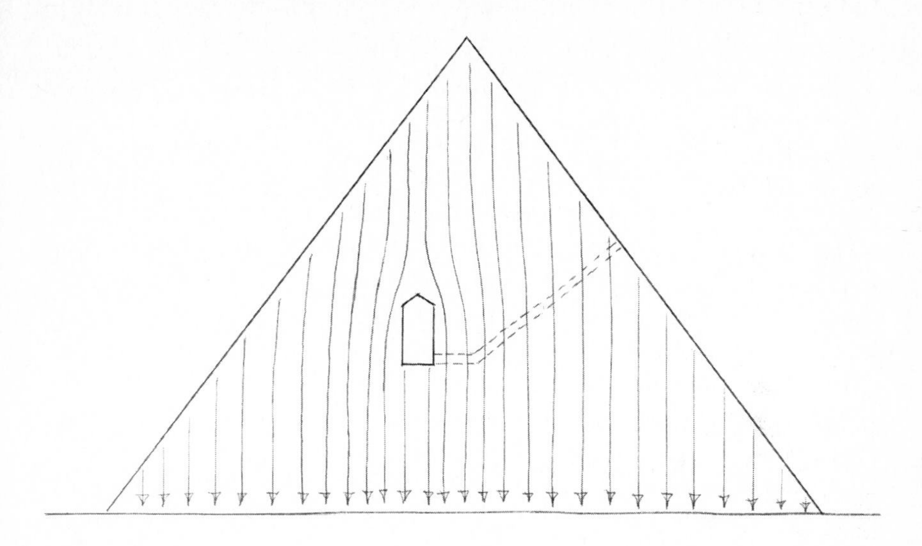

2.5 THE FLOW OF WEIGHT DOWN THE PYRAMID

increases only as the square of the height. Thus, the pressure on the stones at the bottom of the Great Pyramid is more than twice the pressure at the bottom of the Pyramid of Mycerinus.

In addition to the King's chamber, a pyramid had a number of smaller chambers, some destined to contain the alabaster jars for his entrails, another, possibly, to entomb his queen, and others to store his worldly goods, including food and clothing. A complicated set of passages connects these chambers, while other passages were dug for ventilation purposes. It would also seem that some dead-end passages were introduced to mislead the interlopers, even if they had succeeded in lifting or destroying the heavy *portcullis,* the vertical stones sliding into vertical grooves used to block the main passage to the King's chamber after the body of the King had been entombed. As many as three portcullis, one behind the other, were used to stop the thieves, most of the time in vain. A legend developed that the last workers, in charge of lowering the portcullis, were trapped inside the pyramid and sacrificed to the King, but modern research has determined that escape hatches were always provided and that the priests in charge of this last operation did not die in the process.

The erection of a large pyramid was a superb engineering achievement, designed in all its details ahead of time. The granite sarcophagus in the King's chamber of the Great Pyramid, for instance, is so large that it could not have passed through the narrow passages leading to its final location. The sarcophagus was placed inside the pyramid during its construction and the pyramid was actually built around it.

What kind of human effort was required to build such colossal works? We owe to the investigations of Dr. Kurt Mendelssohn, a German physicist and passionate pyramidologist, the estimates of the labor that went into the erection of the main pyramids built during 100 years at Meidum, Dashur, and Gizeh. Between them, these three pyramids required 25 million tons of materials, including limestone, mortar, and brick. Mendelssohn divides the workers into two groups. The first consisted of unskilled masons who quarried the local stone, built the ramps, and lifted the blocks, and of auxiliary workers who lubricated the sledges with water, fed the gangs, and repaired the transport lines. The other group was made up of skilled masons, employed at the Tura quarries and at the pyramid sites to cut, prepare, and set the casing and finished stones. The first group may have been 70,000 strong and was employed seasonally during the floods of the Nile; the second numbered possibly 10,000, permanently employed. Facilities to house 4,000 of them have been found near the pyramid of Chephren. Since Mendelssohn has also convincingly demonstrated that many of the pyramids were built simultaneously, the total number of people employed could have been as high as 150,000, an extraordinary number of workers to be fed and kept by the state in relation to the Egyptian population of 5,000 years ago. The only historical information available on the subject comes from Herodotus. Having talked to the priests of Heliopolis in 400 B.C., he reports that 100,000 men working in three-month shifts labored twenty years to build the Great Pyramid. Since, contrary to the opinion of old investigators, Herodotus has been shown to be quite correct in his explanation of the construction procedures used by the pyramid builders, one is inclined to accept the figures of Mendelssohn as accurate, at least in order of magnitude. The consequences of these estimates, together with those of a fundamental discovery of Mendelssohn, are basic in establishing a fascinating social purpose in pyramid construction, which had not been previously considered by archeologists.

The pyramid at Meidum has always presented a number of unanswered questions stemming from its present shape and some essential differences in its construction and passages. In its present state the pyramid is a ruin, surrounded by 250,000 tons of limestone fallen from it on all four sides. It was formerly thought that, since the pyramids had been used for centuries as quarries, these outer layers of the Meidum pyramid had been stolen to build palaces and bridges in Cairo and other large cities of Egypt. The trouble with this explanation is that there are no large cities in the vicinity of Meidum. Furthermore, most of the "stolen" blocks weigh two tons and are still there. What remains of the pyramid is a three-step structure emerging from a sea of sand and blocks. A careful inspection of this structure reveals three significant features of the original pyramid. The casing at its bottom, still intact, indicates that the 52° angle was used in its erection. The foundation under the casings rests on large blocks of solid stone but is set directly on the desert sand. The blocks of the casing are set in horizontal layers and not in layers inclined inward

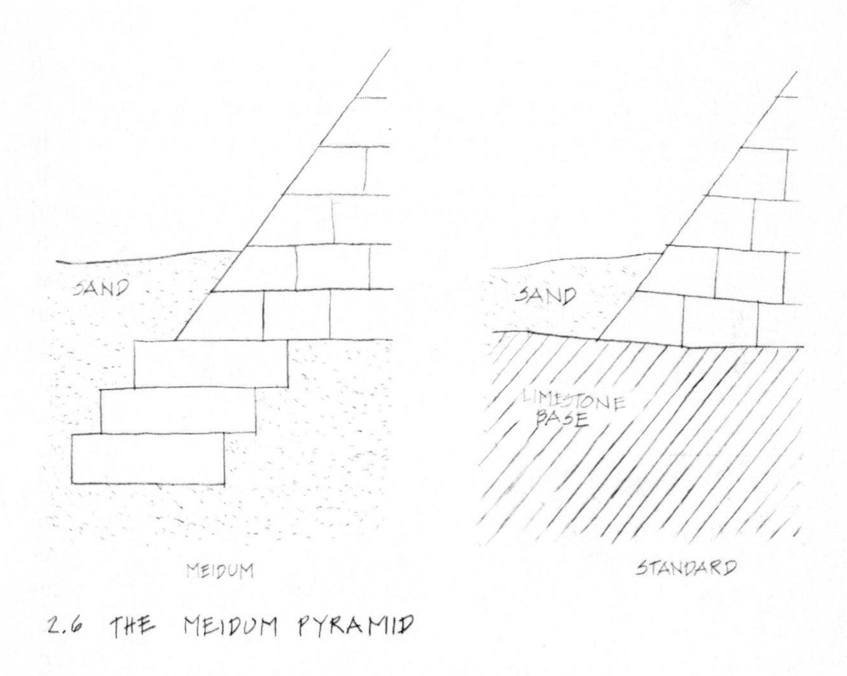

2.6 THE MEIDUM PYRAMID

(Fig. 2.6). It is not difficult to surmise from these three features that the outer layer of the Meidum pyramid was not dismantled by thieves, but collapsed because of structural defects and lies at present at the foot of the existing structure. If this hypothesis is accepted, it becomes easy to understand why the next pyramid to be built after the collapse of the Meidum pyramid, the Bent Pyramid of Dashur (Fig. 2.7), started at a slope of 52°, about halfway up was continued at an angle of 43 ½°. This

2.7 THE BENT PYRAMID AT DASHUR

unique example of change in angle, at first ascribed to the hurry with which the pyramid had to be finished, is much better understood as an architectural lecture well learned: the builders of the Bent Pyramid, warned by the Meidum collapse, decided to continue construction at a safer, smaller slope. A counterproof of this hypothesis is found in the angle of the next pyramid, the Red Pyramid at Dashur, the only pyramid in existence built entirely at the safer angle of 43 ½°.

The engineers of all the later pyramids understood the cause of the Meidum collapse and, having remedied the structural deficiencies that caused it, resumed construction at the classical 52° angle.

The Social Purpose of the Pyramids

As Mendelssohn surmised, more than one pyramid must have been built at the same time during the kingdom of the same Pharaoh. In fact one of them, Seneferu, had certainly two and very probably three pyramids erected in his name. One can only conclude that, since even a Pharaoh could only have one body, some of the pyramids were cenotaphs and not tombs. Why would the priesthood that ruled Egypt for thousands of years go to such extremes of extravagance to honor the Pharaohs? Why would such an onus of taxation and labor conscription be imposed on the people of Egypt? Why did these large masses of laborers, slaves under the whip of harsh masters, accept these conditions at a time when no physical means of total repression was available to the ruling class? The answers given by Mendelssohn to these questions are both historically significant and technically fascinating.

When Menes, the King of southern Upper Egypt conquered northern Lower Egypt, he assumed the name of King of Upper and Lower Egypt, and the administration of the *nomes* or provinces of the new united kingdom became centralized. But it was only after the establishment of the new capital, Memphis, and the acquisition of power by the priesthood at Heliopolis that a state religion developed. No better way of uniting the populations of the two separate kingdoms could be found, after the long conflict between them, than to deify the King. This occurred at the beginning of the Pyramid Age. And what better way of assuring economic equality for the peasants of the forty-two *nomes,* each with their different customs and gods, than to establish a draft in the service of the King-God? The supporters of today's draft stress the economic, cultural, and political advantages of such an institution and the Egyptian peasants might have been ready, perhaps even eager, to serve during the flood period, exchanging their labor for the support from the King's granaries. Thus an army of 150,000 healthy young men became available for the erection of the pyramids. They could not very well be sent home when a pyramid was completed without disrupting the economic and labor balance of the kingdom. Some Pharaohs are known to have reigned for many decades, while others were on the throne for only a few years. It became imperative, then, both to begin the construction of a new pyramid before the death of a king, and to finish the pyramid started by a prematurely dead king while starting the tomb for the new king. In either case work on the pyramids became a continuous process, which welded the nomes into a modern state. All the features of a centralized, powerful, and skillful administration were developed at this time and have remained the model of the state to this day. If this explanation is valid, pyramid construction became one of the most important activities and justifications of the state. One could go so far as to say that the pyramids ran the Pharaohs rather than the Pharaohs running the pyramids. The Mendelssohn hypothesis gives also a rational explanation of the contemporaneous building of pyramids and of the fact that a single King could have one, two or even three pyramids in his name, only one of which could be his actual tomb.

It must be emphasized that the land of Egypt still hides a fund of information about its history and that Egyptologists have barely begun to unravel the complex sequence of events during its 5,000 years. Practically all statements about ancient Egypt are based on interpretations and assumptions which can seldom be verified with absolute certainty. While the Mendelssohn studies seem to be quite factual in terms of the collapse of the Meidum pyramid and the temporary adoption of the 43 ½° slope, his social surmises are less certain, although they may appeal strongly to our modern conception of the state.

The Death of the Pyramids

Pyramid construction lasted for 2,500 years, starting with the *mastabas,* reaching a climax in the IV Dynasty (4,600 years ago) and slowly degenerating into the erection of smaller and smaller monuments of lower quality. Stone blocks, less heavy and haphazardly assembled, became the rule and finally sun-baked bricks were substituted. The guilds of skilled masons vanished and the quarries at Tura were abandoned. The King who succeeded Chephren, Shepseskaf, had the Mastabat Fara'un built for his tomb rather than a pyramid. This indicates a slow transition to new types of worship, requiring a smaller participation of the population in the former basic, ritual activity. The monolithic obelisk became in due time the symbol of the Sun-god Re, although the first obelisks were built of superimposed stones and had a squatter structure, reminiscent of the pyramids.

If the spire of an Egyptian obelisk seems puny compared to the Great Pyramid, let us remember the engineering difficulties of transporting such an obelisk from the back to the front of Saint Peter's in Rome. This process was described in minute detail by Domenico Fontana in 1590. The obelisk, cut from the granite of Upper Egypt in the tenth century B.C., weighed 361 tons. Transportation and erection required a complex set of block-and-tackles and the use of 140 horses and 907 men (Fig. 2.8). The men were given the sacraments, and strict working rules, under penalty of death, were enforced during the operation. Lowering the obelisk took one day, but its transportation along the 800 feet of a specially prepared roadway took six months. Erection at the new site was achieved in one day. The sophisticated engineers of the Italian Renaissance had a hard time dealing with the puny monuments of Egyptian decadence!

The fascination of the Pyramids is due as much to their mystery as to their size. As the creator of the Eiffel Tower, for forty years the tallest tower in the world, put it: "There is an attraction and charm inherent in the colossal that is not subject to ordinary theories of art. Does anyone pretend that the Pyramids have so forcefully gripped the imagination of men through artistic value? What are they but artificial hillocks? And yet what visitor can stand emotionless in their presence?" In the ancient past, religious awe and sun-worship dominated the feeling of the onlooker. Yesterday we admired the Pyramids as extraordinary engineering feats.

2.8 **The Erection of St. Peter's Obelisk (from** Fontana)

Tomorrow we may be moved once again by a social achievement akin to the contemporary construction of enormous dams by the Chinese masses. Today, sensitized to the aesthetic values inherent in elementary geometric forms by an entire school of minimalist sculptors, we may look at the Great Pyramid as the greatest of minimal sculptures. The message changes with the evolution of our cultures, but the mysterious fascination of the Pyramids remains.

3 Loads

The collapse of the Meidum pyramid demonstrates that the problem of *loads,* of weight distribution, even in apparently simple geometric structures, is a complex, ever-present concern for builders. If the earth did not pull, the wind did not blow, the earth's surface did not shake or sink, and the air temperature did not change, loads would not exist and structure would be unnecessary. This would be, indeed, the Alice-in-Wonderland world of architecture where attention could be focused only on the definition and enclosing of interior spaces. But in the real world, the builder must concern himself with structure; structure supports all the loads that act, unavoidably, on buildings. The engineer's first job is to determine which loads will act on a structure and how strong they might be in extreme cases. This is anything but a simple task. Let us look into the world of loads.

Dead Loads

A structure consists of heavy elements like columns, beams, floors, arches, or domes which must, first of all, support their own weight, the so-called *dead load.* And here lies the paradox of structural design. To determine the weight of a structure, once the dimensions of its elements are established and the material chosen, one has only to compute the volume of the elements and multiply it by the weight of a unit volume of the material, its *specific weight.** Tables of specific weights are available to the engineer to facilitate this basic but boring task. The trouble is that, for example, in order to make sure that a beam will carry its own weight (*and* other loads on it), we must first know its dimensions, but these in turn, depend on the beam's weight. Structural design, the determination of the shape and dimensions of structural elements, can only be learned by experience.

The dead load is a load "permanently there." In some structures built of masonry or concrete it is often the heaviest load to be supported by the structure. By the way, any other load also permanently there is always included in the dead load—the weights of the flooring, ceiling, and insulation materials, for example. Similarly, the weight of the *partitions*—the walls dividing one room from another, which may be changed or shifted in rearranging the plan of an apartment but will always be there—must also be included in the dead load.

Live Loads

In addition to its dead load, a structure must support a variety of other weights—people, furniture, equipment, stored goods. These impermanent or *live loads* may be shifted around and they may change in value. One may be alone in one's room today and have ten visitors tomorrow. These may gather in one corner or spread themselves throughout the room. The next tenant may have massive furniture and distribute it differently. It is obvious that we can never know exactly what the live load is and how it is going to be distributed.

Concern for safety suggests that live loads must be established on the basis of the *worst* loading conditions one may expect during the entire life of the structure. These are determined by responsible and practiced engineers and contained in *building codes,* which are published by cities, counties, and states. In the United States a few codes have gained general acceptance and most local codes are based on their prescriptions. For example, the so-called Uniform Building Code is akin to the engineer's Bible and no designer ignores the live load values suggested in it.

The values of the code loads are conventional. They assume that the worst effect of the varying and shifting live loads may be represented

* For example, a reinforced concrete beam 30 feet long, 1 foot wide, and 3 feet deep has a volume of $30 \times 1 \times 3 = 90$ cubic feet. Since one cubic foot of concrete weighs 150 pounds, the dead load of the beam is $150 \times 90 = 13{,}500$ pounds.

by a *uniform load,* that is, a load evenly spread over the surface of the floors. For example, the New York City Code suggests that the live-load allowance for a private apartment room should be forty pounds per each square foot of floor. (By the way, engineers abbreviate the words "pounds per square foot" as "psf.") This is a most conservative allowance, but engineers *must* be conservative when confronted with the uncertainties of live loads. It is better to run scared than to be responsible for a failure which may damage property and even kill people.

On the other hand, the chances are absolutely minimal that each square foot of each room at each floor of a building will be loaded *at the same time* by the full code allowance, and buildings designed for this absurd assumption would become unjustifiably expensive. Hence, the codes allow a *live load reduction,* which may reach sixty percent for a high-rise building.

Naturally, the value of the live load varies with the type of building, its location, and its importance. The floors of a warehouse must be expected to carry a much greater live load than those of an apartment house. The roof of a building in Colorado must support a much heavier load of snow than one in Alabama. The public areas of a building, its corridors and halls, which at times may be jammed with people, must be designed for larger live loads than a private room.

Live-load calculations are lengthy and important, although not very demanding on the imagination and intelligence of the engineer. Luckily, computer programs do most of the evaluations now, saving engineering time and increasing both speed of calculation and accuracy of results. The computer, when properly used, is a wonderfully useful slave.

Dynamic Loads

The dead load is permanent and unchanging and the live loads have been tacitly assumed to change slowly, if at all. Together, these unchanging or slowly changing loads are called by engineers *static loads,* loads that stay.

But other loads change value rapidly and even abruptly, like the pressure of a wind gust, or the action of an object dropped on the floor. Such loads are called *dynamic* and may be exceedingly dangerous because they often have a much greater effect than the same loads applied slowly. For example, a hammer laid slowly, gently, onto the head of a nail will make no impact. But dropping the same hammer suddenly on the nail will drive the nail into the wood. Such suddenly applied loads, actually called *impact loads,* can be shown to be fleetingly equivalent to many times their statically applied weights. The dynamic pressure of a slap, compared to the static pressure of a caress, may never have been measured, but one can certainly feel it.

The fleeting effects of a dynamic load depend on how fast the load varies. This raises a question: for example, should the pressure on a building created by a wind gust, first increasing and then decreasing, be considered a static or a dynamic load? The answer is that no varying load is ever static or dynamic *in itself*. As we shall see, its effects can be static or dynamic *depending on the structure to which it is applied.* To prove this let us consider a tall building acted upon by a wind gust.

Under the wind pressure the building bends slightly and its top moves (Fig. 3.1). Its movement may be small enough not to be seen by the naked eye, or even sensed, but since structural materials are never totally rigid, all buildings do sway in the wind. If one could push the top of a building, say, one foot to the right and then let it go, the building would start oscillating, going back and forth. Its top would first go back through its original vertical position, then move one foot to the left of it,

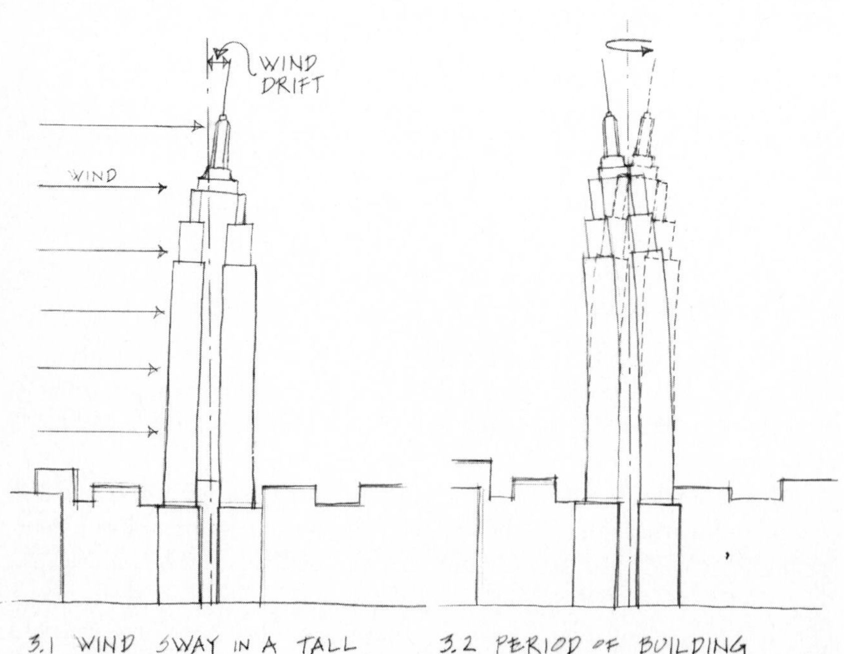

3.1 WIND SWAY IN A TALL BUILDING

3.2 PERIOD OF BUILDING OSCILLATIONS

and continue swinging back and forth until it eventually stops. It is easy to visualize these oscillations by considering the building as an up-side-down clock pendulum, which also swings back and forth when displaced from its lowest position. The time it takes a pendulum to complete a full swing, from extreme right to extreme left *and back*, is called the *period* of the pendulum. Similarly, the time it takes a building to swing through a complete oscillation (Fig. 3.2) is called its period. For example, the period of the oscillations of the steel towers of the World Trade Center in New York City, which are 1,350 feet high, is ten seconds, while the period of a ten-story brick building may be as short as half a second.

We can now answer the question about the effect of wind pressure on a building. The action of the gust depends not only on how long it takes to reach its maximum value and decrease again, but on the period of the building on which it acts. If the wind load grows to its maximum value and vanishes in a time *much shorter* than the period of the building, its effects are *dynamic*. They are *static* if the load grows and vanishes in a time *much longer* than the period of the building. For example, a wind gust growing to its strongest pressure and decreasing in two seconds is a dynamic load for the World Trade Center towers with a period of ten seconds, but the same two-second gust is a static load for the ten-story brick building with a period of only half a second. In a sense, a force the building can slowly absorb is static; an unexpected one is dynamic. The weight of snow and people is always a static load because snow takes hours to accumulate and people enter buildings singly or in small groups. On the other hand, the explosion of an atomic device reaches its maximum effect and decreases so rapidly (less than a thousandth of a second) that it is a dynamic load on all structures, and has enormously destructive effects.

Interestingly enough, there are loads which, though not growing rapidly, do have dynamic effects increasing, not instantaneously, but progressively in time. This phenomenon, called *resonance*, is one of the most dangerous a structure may be subjected to. To understand resonance, let us consider how a heavy church bell, which swings like a pendulum, is made to ring by the relatively small yanks of a single man on its rope. If the bell weighs a few tons—often the case—the ringer might try in vain to move it with a single yank. But if the ringer starts yanking the rope with a small pull of, say, a few pounds and before yanking it again, waits for the bell to go through its first tiny swing, then keeps yanking in *rhythm* with the bell's oscillations, eventually the bell swings widely and rings. The trick here consists in yanking the rope at the beginning of each new oscillation, that is, at time intervals equal to the period of the bell, so the applied pulls will add up.

When a force is rhythmically applied to a structure with the same period as that of the structure, the force is said to be *in resonance* with it. Resonant forces do not produce large effects immediately, as impact forces do, but their effects increase steadily with time and may become catastrophic if they last long enough. If a long series of wind gusts, growing and waning in pressure with a relatively slow period of ten seconds, were to hit the World Trade Center towers, the swing of the towers would slowly increase until the structure of the building might sway so widely as to collapse. The story is often told of a German army infantry company goose-stepping across a small wooden bridge in rhythm with the period of the up-and-down oscillations of the bridge. The company ended up in the river when the bridge collapsed under the resonant load of the goose-step.

There are, finally, some perfectly steady forces which produce dynamic effects on certain types of structures. They derive from the interaction between the wind and the structure and are called *aerodynamic*. In 1940 the Tacoma Narrows Bridge in Washington, a particularly narrow and flexible suspension bridge 2,800 feet long, was destroyed by a steady wind blowing at forty-two miles per hour for about seventy minutes. Up-and-down oscillations of the bridge, travelling like a wave along the length of its roadway, had been noticed since its construction. Indeed, the bridge had been nicknamed "Galloping Gertie." These oscillations were caused by winds blowing at right angles to the bridge which, as the flexible roadway moved up and down, hit the bridge alternatively from below and from above, naturally in resonance with the period of the bridge oscillations. Their effects, while similar to those of a resonant load, had never been sufficiently strong to wreck the bridge. The collapse occurred when similar, but twisting, oscillations were added to the roadway by the wind blowing not horizontally, but at a slight downward angle. The downward wind pressure first pushed down slightly the *windward* edge of the roadway, twisting it; then the structure, reacting, twisted the roadway back up, thus allowing the wind to push the windward edge up from below. This cycle was repeated, again and again, gradually increasing the twisting oscillations of the roadway until the wind destroyed the bridge. Luckily nobody was killed, since authorities had cleared the bridge. The roadways of all modern suspension bridges are now stiffened against twisting to prevent this dangerous phenomenon.

It is interesting to notice that, although the recent technical literature had never considered the aerodynamic effects of winds on suspension bridges, English newspapers at the beginning of the nineteenth century carried descriptions of the collapse of flexible suspension bridges caused by the identical aerodynamic phenomenon that destroyed the Tacoma Narrows Bridge. In the history of science and engineering, facts and laws have been forgotten which, if remembered, would have saved time, energy and, possibly, lives.

Wind Loads

The forces exerted by winds on buildings have dramatically increased in importance with the increase in building heights. Static wind effects rise as the square of a structure's height and the high-rise buildings of the 1970s, which are at times almost 1,500 feet tall, must be fifty times stronger against wind than the typical 200-foot buildings of the 1940s. Moreover, the speed of wind grows with height, and wind pressures increase as the square of the wind speed. Thus, the wind effects on a building are compounded as its height increases.

Wind pressures act horizontally and, in tall buildings, require a structure separate and different from that which resists the vertical gravity loads—the weights. In very tall buildings up to ten percent of the structural weight, and hence of the structural cost, goes into this *wind bracing*. In a twenty- or thirty-story building, on the other hand, the gravity structure is often sufficient to resist the wind. The chances are slim that the strongest possible wind will occur when the structure is also loaded by the heaviest possible gravity loads. Thus, the codes allow structural materials to be stressed thirty-three percent more when the loads due to gravity and those due to wind are taken into account simultaneously. This judicious allowance, based on probability, allows considerable structural savings without impairing structural safety.

One of the basic questions to be resolved before designing a building is often: "What is the strongest wind to be expected at its site?" To answer it, wind measurements are conducted daily in all parts of the world and maps are plotted, like that in Figure 3.3, which give at a glance the maximum wind speeds to be expected. But should one design for the strongest wind ever at the site of a building? If this maximum wind were always taken into account, the cost of the wind structure would become unjustifiably high. It is wiser to design buildings so that they will be undamaged by a wind with a chance of occurring once in, say, 50 years, but to allow *minor* damage under the forces of a *100 year wind* (as a wind is called that has a chance of occurring once in 100 years). The cost of fixing the minor damage would certainly be less than the long-term cost

3.3 WIND VELOCITY MAP OF THE UNITED STATES (MILES PER HOUR)

required to guarantee total integrity under the stronger wind. It is thus up to the engineer to choose the *design wind* and a 30 or a 50 year wind is often considered acceptable unless the strength of a 100 year wind is such as to endanger the entire building and the life of its occupants.

Besides depending on wind speed and building height, wind forces vary with the shape of the building. The wind exerts a pressure on the windward face of a rectangular building because the movement of the air particles is stopped by this face. The air particles, forced from their original direction, go around the building in order to continue their flow, and get together again behind the building as shown in Figure 3.4. In so doing, the air particles suck on the leeward face of the building and a negative pressure or *suction* is exerted on it. The total wind force is the sum of the windward pressure and the leeward suction, but each of these two forces has its own local effects. During hurricane Donna in New York City, in 1960, occupants of high-rise buildings were justifiably frightened when large glass panels of the curtain walls were blown *into* their offices by the wind pressure. They were probably even more frightened and amazed when the leeward wind action sucked the window panels *out* of their offices. These leeward panels fell to the street, creating an additional hazard to passersby.

In designing for wind, a building cannot be considered independent of its surroundings. The influence of nearby buildings and of the land configuration can be substantial. The buildings around a new hall at M.I.T. in Cambridge, Massachusetts, produced such an increase in wind velocity at ground level that it became, at times, impossible to open the entrance doors. Originally the two curved structures of Toronto's City Assembly Hall had only a narrow space separating them. This defile compelled the wind to increase its speed in order to pass through it. (Air is a fluid and increases its speed when constrained to go through a narrow space, just as water increases its speed when passing through a narrow gorge of a river.) This increase in speed created such high pressures on the two buildings that their wind bracing had to be redesigned on the basis of information gathered in wind tunnel tests.

The swaying of the top of a building due to wind may not be seen by the passerby, but it may feel substantial to those who occupy the top stories of a high building. Under a strong wind the tops of the World Trade Center towers swing left and right of their vertical position by as much as three feet, and a hurricane can produce swings of six to seven feet on each side of the vertical. These horizontal swings are not structurally dangerous, but they may be inconvenient for those who work at such great heights: occupants sometime become airsick. Recent research by engineers and doctors indicates that the airsickness induced by wind motion in high buildings is a resonance phenomenon. It occurs when the period of the building more or less coincides with the period of the up-

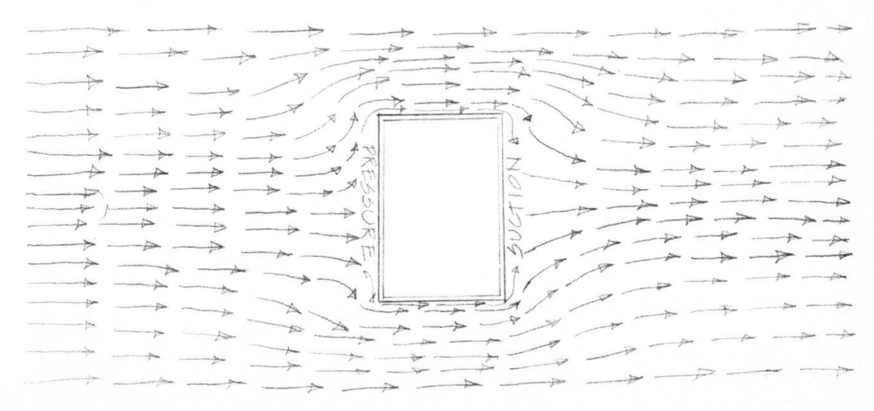

3.4 WIND FLOW PAST A BUILDING

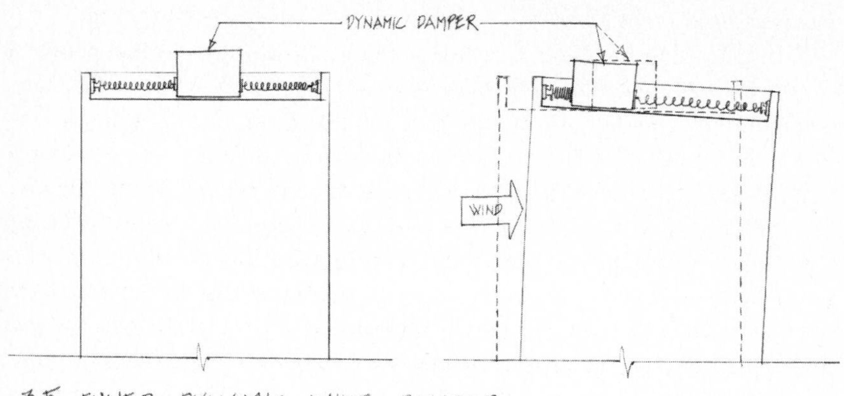

3.5 TUNED DYNAMIC WIND DAMPER

and-down oscillations of our own insides. This explains why some, but not all, of the people in a building may feel queasy. To avoid excessive wind deflections (or *wind drift* as it is technically called) buildings should be stiffened so that their tops will never swing more than 1/500 of their height. Thus a three-foot wind drift is acceptable in a 1,500 foot building.

Lateral movement due to wind may be even more dramatic in long and flexible suspension bridges. After the collapse of the Tacoma Narrows Bridge, the actual lateral swings of all large suspension bridges in the United States were measured. It was found, for example, that the Golden Gate Bridge in San Francisco sways laterally as much as eleven feet under heavy winds. Sometimes the bridge must be closed to traffic since it is not safe to drive a speedy car on a roadway that moves right and left under it.*

How does one prevent the resonant oscillations in a building? The basic method consists in changing its period by reinforcing its structure to make it stiffer. The stiffer the structure, the shorter the period. This is

* Tall chimneys and high television towers, which have reached over 2,000 feet and are among the highest structures ever built, are subjected to a dangerous resonance phenomenon, in which they oscillate *at right angles* to the wind. As the wind hits these extremely slender structures, it tends to go around them and creates air eddies which separate from the sides of the structure, causing an alternating vacuum, first on one side and then on the other. When the air eddies separate with a rhythm that coincides with the period of the chimney or tower, the structure starts oscillating laterally with increasing swings, which may damage or eventually destroy it.

a costly remedy but recently a mechanical gadget used for years in cars and planes has been adapted to the reduction of wind oscillations. It is called a *tuned dynamic damper* and consists of a heavy mass of concrete attached to the top of the building by means of lateral springs (Fig. 3.5). This heavily springed mass has the same period as the building. When the building oscillates with its own period, the tuned damper after a short while also tends to oscillate with the same period, but in the *opposite* direction. One could say that the damper moves in antiresonance with the building. When this happens, the oscillations of the building are completely damped out by the counteraction of the damper. The damper's resonant oscillations do not grow because they are controlled by large shock-absorbers that brake its motion. A spring-connected, 400-ton mass of concrete at the top of the Citicorp Building in New York City reduces its top oscillations by 50% without a substantial increase in the cost of its wind-bracing. Similar dynamic dampers, consisting of heavy metal rings connected to the top of chimneys by radial springs, are used to avoid the lateral swings of these high structures due to air eddies.

We have acquired a great amount of knowledge about wind in the last twenty or thirty years and it is comforting to know that very few modern well-designed structures suffer today from wind damage.

Earthquake Loads

Earthquakes have wreaked destruction since oldest antiquity and it is only in the last thirty or forty years that our knowledge of earthquakes and of their impact on buildings has resulted in the design of earthquake-resistant structures. These are built with particularly strong "wind bracing" type structures, which tests and computer calculations prove capable of resisting the jerking forces of an earthquake. Even so, the number of quake victims is still high all over the world. When 27,000 people died in the Guatemala earthquake of 1967 we thought we had seen the worst, but 242,000 people died in an earthquake later in the same year in the region north of Peking.

The earth's crust floats over a core of molten rock and some of its parts have a tendency to move with respect to one another. This movement creates stresses in the crust, which may break out along fractures called *faults*. The break occurs through a sudden sliding motion in the direction of the fault and jerks the buildings in the area. Since the dynamic impact forces due to this jerky motion are mostly horizontal, they can be resisted by the same kind of bracing used against wind.

Earthquake strengths are evaluated on scales like the Richter scale, which measures the magnitude of the energy in the earthquake. For example, an earthquake measuring 4 or 5 on the Richter scale does little damage to well-built buildings, while one measuring 8 or above collapses buildings and may cause many deaths. Not all parts of the earth are subjected to earthquakes, but there are two wide zones on the earth's surface where the worst earthquakes take place. One follows a line through the Mediterranean, Asia Minor, the Himalayas, and the East Indies, the other the western, northern, and eastern shores of the Pacific.

It is comforting to know that we are on the threshold of accurate earthquake forecasting and that some earthquakes have already been fairly well predicted in the United States and abroad. (In this we lag behind horses and cows which give obvious signs of fear at the approach of an earthquake.) Our new capabilities are due to the fact that when an explosion is detonated at a point in the earth's crust, waves move out concentrically through it, as they do when we drop a stone in the calm waters of a lake. It has been proved that the wave-speed through the crust increases when the stresses in it increase. Thus, when a geologist notices that the velocity of the waves created by a small explosion increases, he knows that the stresses in the crust are increasing and concludes that an earthquake may be imminent. The first successful prediction in the United States was made by a Columbia University student in the Adirondacks only a few years ago.*

Thermal and Settlement Loads

The last category of loads the engineer must worry about consists of those caused by daily or seasonal change in air temperature or by uneven settlement of the soil under a building. These are sometimes called hidden or *locked-in loads.*

Let us assume that a steel bridge 300 feet long was erected in winter at an average temperature of 35°F. On a summer day, when the air temperature reaches 90°F., the bridge lengthens, since all bodies expand

* While earthquake prediction will certainly be a great boon to humanity, it will present extremely complex social problems. What will the mayor of Los Angeles do if he is told by scientists that a strong earthquake will hit the city a few days or even hours hence? Is it possible to evacuate at short notice a city of 7 million people in an orderly manner? What if at first the geologists are not accurate and predict an earthquake that does not take place? Will people believe them if they cry wolf once too often? Scientific progress can only be a blessing to humanity if man learns to use its results wisely.

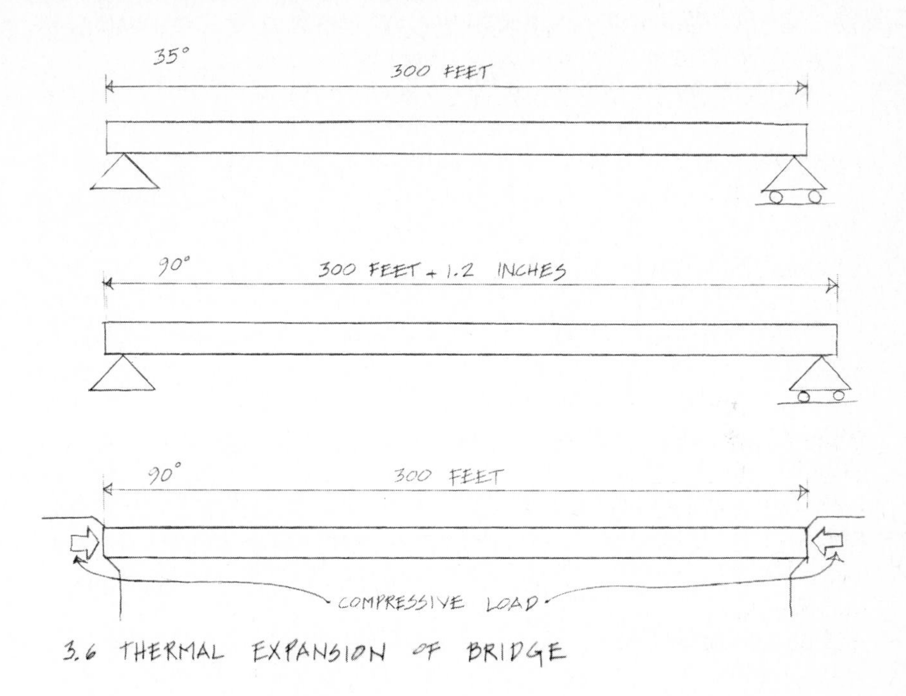

3.6 THERMAL EXPANSION OF BRIDGE

when heated. The increase in length of the bridge can be computed to be only 1.2 inches (Fig. 3.6). It is indeed small, one three-thousandths of the bridge length, but, if the bridge is anchored to abutments that do not allow this thermal expansion, the abutments will push on the bridge to reduce its length by 1.2 inches. Unfortunately, steel is so stiff that the compressive load exerted by the abutments uses up half the strength of the steel. There is only one way of avoiding this dangerous overstress: one of the bridge ends must be allowed to move to permit the thermal expansion to occur. While gravity loads must be fought by increasing the strength and stiffness of a structure, thermal loads must be avoided by making the structure less rigid.

Similar thermal loads appear at the bottom of large domes, which tend to expand when the temperature of the ambient air increases. Due to thermal expansion, the lower edge of a large dome tends to move outward and, since it is impractical to allow this edge to move in and out depending on temperature changes, one must build around it a strong ring to prevent this motion. Most of the large domes built in the past have shown a tendency to crack at their supported boundary (due to both thermal and gravity loads) and have since been circled with hoops of steel. Hagia Sophia in Constantinople and Saint Peter's in Rome have been so reinforced after cracks appeared at their edges.

In many modern high-rise buildings the structural frame is set outside the curtain wall, rather than hidden inside it. An emphasis on the importance of structure and a feeling for its aesthetic value is at the basis of this architectural innovation. Good examples are the 100-story John Hancock Insurance Company steel building in Chicago (see Fig. 7.9) and the Columbia Broadcasting System concrete building in New York (see Fig. 7.12). While there is no denying the aesthetic value of some of these exposed structures, it must be noticed that they create problems for the engineer. The interior of these buildings is air conditioned and maintained at a constant temperature of 65° to 72°F., while the exposed structure is subjected to air temperature changes. In summer the exterior columns may reach a temperature of up to 120°F., and become two or three inches longer than the interior columns, while in winter they may become that much shorter when their temperature goes down to 20°F. These variations in length do not damage the columns, but, as shown in Figure 3.7, they bend the beams connecting the outer to the inner columns, particularly at the higher floors. These would be badly damaged if they were not properly designed, either by reinforcing them or by

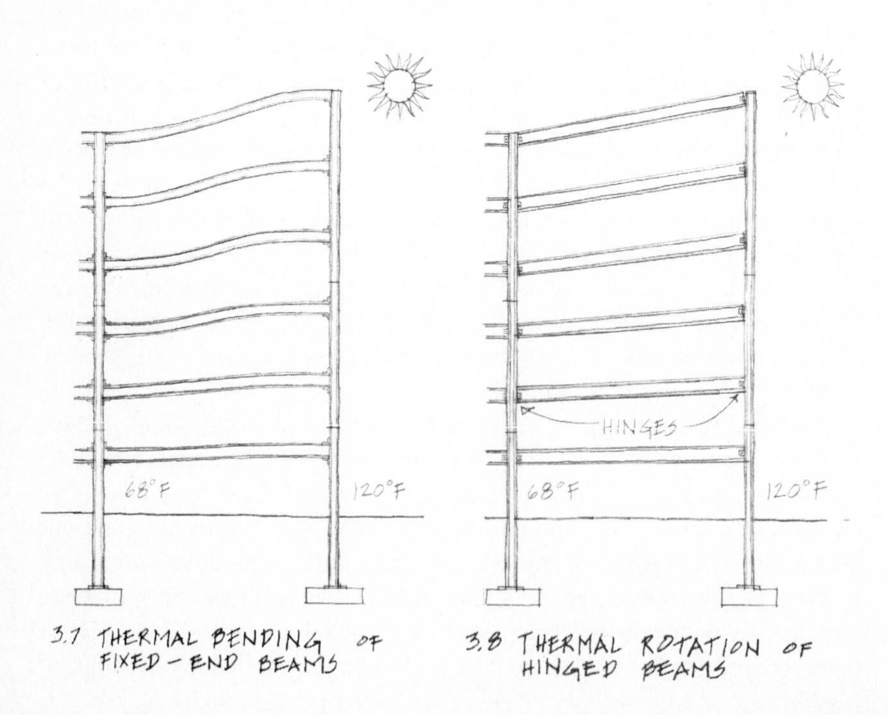

3.7 THERMAL BENDING OF FIXED-END BEAMS

3.8 THERMAL ROTATION OF HINGED BEAMS

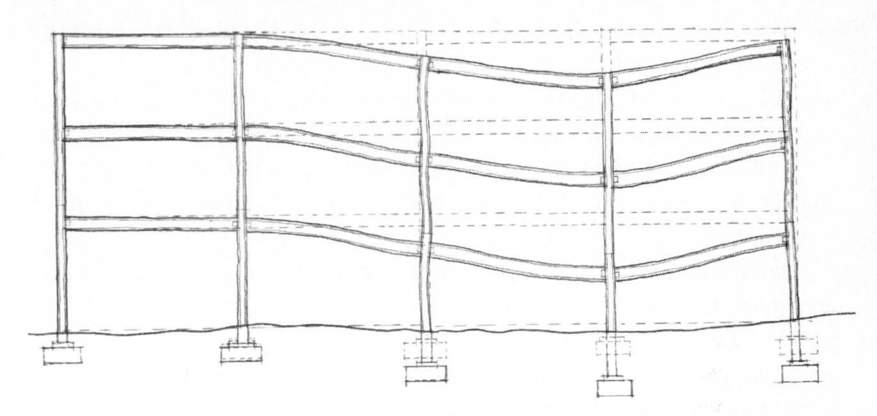

3.9 UNEVEN SETTLEMENT OF BUILDING FOUNDATIONS

allowing their ends to rotate, that is, by *hinging* them to the columns (Fig. 3.8).

It is easy to conceive of a mechanical way of avoiding these beam stresses: have the hot and cold water systems circulate inside the outer columns of the building and regulate the water flow so that the temperature of the outer columns be always more or less equal to that of the inner columns. This system has not been applied yet because of its cost, but the United States Steel Building in Pittsburgh does have a cold-water circulation system in its outer columns. Its purpose is to prevent any heat generated at a point on the outer structure, say, by a fire, from dangerously heating and possibly melting these columns.

Bending of the beams connecting outer to inner columns may also occur if the soil under a building settles unevenly (Fig. 3.9). Such uneven settlements caused the leaning of the Tower of Pisa, which started while the tower was being built. The Pisans thought they had straightened the tower up by building its upper part vertically, but the 191-foot tower is still going over at the rate of about one inch every eight years and its top is now out of plumb by sixteen feet. Various measures are being studied to arrest this dangerous rotation, but the Pisans are resolved to stabilize the tower in a leaning position, since no tourist would travel far to look at a straightened-up tower.

It must be emphasized that most damage to buildings is caused by foundation problems. Soil mechanics, the study of soil behavior, has moved to a science from an art only during the last fifty years. The island of Manhattan is blessed by a rocky soil, which permitted in 1913 the erection of the first high-rise (the Woolworth Building). Mexico City, on the other hand, is built on a mixture of sand and water. Such soils settle when heavy buildings are erected, squeezing the water out of the sand. The National Theatre in the center of Mexico City, originally built at grade level with a heavy cladding of stone, in a few years sank as much as ten feet. Downward stairs had to be built to its entrance. People were amazed when later on the theatre began to rise again, requiring the construction of an upward staircase. This strange phenomenon can be explained by the large number of high-rise buildings which had been erected nearby. The water squeezed out from under them by their weight pushed the theatre up.

Loads may be a necessary evil to both the architect and the engineer, but their basic importance cannot be minimized. Wise is the engineer who gives them his care and attention before starting the design of a building.

4 Materials

Tension and Compression

The purpose of structure is to channel the loads on the building to the ground. This action is similar to that of water flowing down a network of pipes; columns, beams, cables, arches, and other structural elements act as pipes for the flow of the loads. Obviously, this becomes a complex function when the structure is large and the loads numerous.

The remarkable, inherent simplicity of nature (Einstein called it elegance) allows the structure to perform its task through two elementary actions only: pulling and pushing. Many and varied as the loads may be and geometrically complicated as the structure may be, its elements never develop any other kind of action. They are either pulled by the loads, and then they stretch, or are pushed, and then they shorten. In structural language, the loads are said to *stress* the structure, which *strains* under stress. The imagery of this terminology is significantly human. When a structure is "overstressed," it "breaks down" and sometimes "buckles." (As will be seen later, this is said of thin elements subjected to compressive loads too high for their capacity.)

Another basic law of nature governs the structure's response to the loads. With a judicious sense of economy, or intelligent laziness, a structure will always choose to channel its loads to the ground by the *easiest* of the many paths available. This is the path requiring the minimum amount of work on the part of the structural materials and is a consequence of what is termed in physics "the law of least work." Structure behaves humanly in this respect too.

When a material is pulled, it is said to be in *tension*. Tension is easy to recognize because it lengthens the material. The cables of an elevator are pulled by the weight of the cabin and the passengers; the cables are longer when they lift or lower the cabin than when they were originally installed. We can detect tension easily in elements made of very elastic materials, like a rubber band. Pull on a rubber band and it easily becomes twice its original length.

When a material is pushed it is said to be in *compression*. Compression, in a sense, is the opposite of tension, since it shortens the material. If we push on a rectangular sponge, the sponge becomes shorter.

Structural materials are much stiffer than rubber bands or sponges. Their lengthening under tension or shortening under compression may not be seen by the naked eye, but it always occurs, since there are no *perfectly rigid* structural materials. As a high-rise building is erected, its lower columns are compressed more and more by the weight of its growing number of floors, but we cannot see the shortening of its columns because, whether they be made of steel or concrete, this reduction in height may amount to only slightly more than one inch in a building 1,000 feet high. The cables of a suspension bridge are set in tension by the weight of the roadway and of the vehicles traveling on it. Their elongation is 1/300 of their length, but cables are prestretched (stretched before being loaded) to reduce it. The tiny changes in length due to tension and compression when divided by the original length of the element are called *strains*. If an elevator cable originally 300 feet long (3,600 inches) becomes 4 inches longer under load, its strain is 4/3,600 or 1/900. The pull or push on an element, divided by the area it is applied to, is called *stress*. If a column 10 inches square carries a load of 120,000 pounds, its stress is $120{,}000 \div (10 \times 10) = 120{,}000 \div 100 = 1{,}200$ pounds per square inch (abbreviated 1,200 psi).

Since all structural actions consist of tension and/or compression, all structural materials must be strong in one or both. *Strength* has quite different values in wood, reinforced concrete, and steel, but all three materials have the capacity to resist tension and compression, that is, to be pulled or pushed by larger or smaller forces before they break under load.

Elasticity and Plasticity

Strength is not the only property required of all structural materials. Whether the loads act on a structure permanently, intermittently, or only briefly, the lengthening and shortening of its elements must not increase *indefinitely* and must *disappear* when the action of the loads ends. The first condition guarantees that the material will not stretch or shorten so much that it will eventually break under the working loads. The second insures that the material and, hence, the structure will return to its original shape when unloaded. Whenever we walk on a wooden footbridge, we notice that its planks move down, even if minimally, under the action of our weight. If this downward displacement were not to disappear when we leave the bridge, the next time we walk on it a second displacement would be added to the first and the floor would sag more and more until it would become unusable. If the lateral displacement of the top of a tall building (its wind *drift*) were not to disappear when a gust stops pushing on it, each new gust would displace the top more and more until the building would look like a gigantic Leaning Tower of Pisa.

A material whose change in shape vanishes rapidly when the loads on it disappear is said to behave *elastically*. A rubber band is correctly called an elastic, since it returns to its length when we stop pulling on it. All structural materials must be elastic to a certain extent, although none is perfectly elastic under high loads.

Most structural materials not only behave elastically, but within limits show deformations that increase in proportion to the loads. Let us say that if a man stands on the tip of an elastic diving board, the tip moves down by one inch, and that, if a second man of equal weight also stands there, the tip deflection becomes two inches. In this case, the diving board is not only elastic, since its deflection vanishes when the weight of the divers disappears, but also *linearly elastic.* (This terminology derives from the fact that if we plot the board deflections versus the loads, the graph we obtain is a straight line.)

The discovery of the linearly elastic behavior of materials, called for short *elasticity*, was made towards the end of the seventeenth century by an Englishman, Robert Hooke, who was a physicist, a surveyor, and an architect. He printed his discovery as an anagram of the statement of his law in Latin: CEIIINOSSSTUU, which he later unscrambled to read: "Ut tensio sic vis" ("As the elongation, so is the force"). His secrecy was motivated by the fear that someone else might patent his discovery in connection with watch springs, which are, of course, typically elastic.*

* Such fears were and are quite common. When the sixteenth-century mathematician Tartaglia discovered the formula for the solution of the cubic equation, he kept it secret until one of his pupils betrayed him and gave the formula to Cardan, who to this day is credited with it.

All structural materials behave elastically if the loads are kept within given limited values. When the loads grow above these values, materials develop deformations larger and no longer proportional to the loads. These deformations, which do not disappear upon unloading, are called *permanent* or *residual* deformations. When this happens, the material is said to behave *plastically.** If the loads keep increasing after the appearance of plastic behavior, materials soon fail.

It has been pointed out that plastic behavior, and hence permanent deformations, is to be avoided in order not to impair the function of a structure. But it must not be thought that plasticity is entirely detrimental in structural materials. On the contrary, it can be quite useful. For example, if we progressively load a structure and measure its increasing deformations, we are warned that the structure is in danger of collapse as soon as we notice that these deformations grow faster than the loads. In other words, materials that behave elastically under relatively small loads and plastically under higher loads do not reach their breaking points suddenly. Once they stop behaving elastically, they keep stretching (or shortening) under increasing loads until they continue to do so even *without* an increase in the loads. Only then they fail. If a steel wire is weighted heavily enough, it will keep stretching or *yielding* under a constant load. It thus gives warning of its impending failure.

Materials which do not yield are called *brittle* and cannot be used in structures, because they behave elastically up to their breaking point and fail suddenly without any warning. This is one reason why glass cannot be used structurally although it is stronger than steel in tension or compression. Thus, strength, elasticity, *and* plasticity are all necessary to good structural behavior.

The value of the loads under which a structure will start behaving plastically depends on a number of conditions, most importantly, temperature. Steel, the strongest structural material available to man, becomes plastic at high temperatures and loses its strength at 1,200°F. Steel buildings must be fireproofed to retard the heating of its columns and beams in a fire. Concrete, instead, is a particularly good insulating material and prevents for a long time the heating and yielding of its reinforcing steel bars. Reinforced concrete buildings do not have to be fireproofed. On the other hand, at a temperature of minus 30°F., called its *transition*

* The word plastic has two different meanings in structures and in chemistry. In structures it indicates the inelastic behavior just described; in chemistry it describes a large variety of man-made organic materials like celluloid or bakelite, some of which, notwithstanding their name, behave elastically.

temperature, steel becomes brittle and breaks suddenly, particularly under impact, or suddenly increased, loads. A steel railroad bridge in northern Canada had its girders break as if made of glass when a train driven by a steam locomotive ventured onto it in the middle of a particularly cold night. A concrete bridge would not have behaved the same way.

Safety Factors

The strength of a structural material is measured by the number of pounds each square inch of material will carry before it breaks. This number, similar to those measuring stress, is called its *ultimate strength* and varies from material to material and even in the same material depending on how it is stressed.

Steel, aluminum and other metals have the same strength whether pulled or pushed. A compressed steel column or a tensioned steel wire can support anywhere from 36,000 to 60,000 pounds per square inch before failing. Aluminum columns and wires behave identically, but deform three times as much as steel. This high deformability of aluminum together with its high cost per pound—large amounts of electricity are needed to manufacture it—make structural aluminum alloys less popular than steel, although its ore (bauxite) is common everywhere. Steel also has the valuable property of changing from elastic to plastic behavior at a well-defined value of the stress, called its *yield stress.* Since plastic permanent deflections must be avoided, we can load steel up to only a fraction of its yield stress, usually sixty percent of it. This reduction ensures safety, and forces, as we have seen, the conservative estimates of the code loads. Due to all these causes, *factors of safety,* which must honestly be considered "factors of ignorance," as high as 5 are used in designing structures. This means that such structures could be overloaded 5 times before collapsing.*

A column of marble, concrete, or limestone could be as high as 12,000 feet before collapsing in compression under its own weight. Stronger

* Whenever a material has a clearly defined stress at which it starts yielding, the factor of safety is established to allow a stress which is a fraction of its yield stress. For example, we have seen that in steel this fraction is 60/100. Hence, the safety factor is only 100/60 = 1.6.

When a material, like concrete, yields under low stress, but does not break soon thereafter, the allowable stress is established as a fraction of its ultimate strength at which it fails. For example, concrete with an ultimate strength in compression of 5,000 pounds per square inch may have an allowable stress of only 2,000 psi, i.e., a safety factor of 2.5.

stone, like granite, could reach 18,000 feet. But the tallest stone buildings ever erected, the limestone pyramids of Egypt, are less than 500 feet tall and, hence, have a factor of safety of at least 24. The Egyptians were very prudent. Unfortunately, stone and concrete resist pulling badly. They can only be used in elements that are never pulled, such as columns or arches, but are not well suited for beams, which, as we shall see, develop both tension and compression.

Wood's strength not only differs in tension and compression, but depends on whether we stress it in the direction of its grain or at right angles to it. To overcome this peculiarity sheets of wood with fibers oriented in different directions are assembled with plastic glues—plywood and laminated woods. These have approximately the same strength properties in all directions. Wood is one of the few natural materials with a high tensile resistance and has been used throughout history in beams and other elements developing tension. Unfortunately it burns at low temperatures and must be fire retarded, even though fire retardation decreases its strength.

Steel

Man-made materials have permitted the realization of the record-breaking structures of modern times. The tallest building in the world today (the Sears Tower in Chicago) is supported by steel columns and beams, and the longest suspension bridge (the Humber Bridge in Great Britain) is held up by steel cables. Pound per pound, steel is the material with the greatest strength obtainable at the lowest price.

Steel is an alloy of iron and carbon, with very tiny amounts of other metals to give it particular properties. For example, nickel is added to make stainless steel, which does not rust. The carbon content of *steel* varies from 0.1 percent to not more than 1.7 percent. Iron with less carbon is called *wrought iron* and is workable at low temperature. The first wrought-iron man-made object is a dagger manufactured in Egypt in 1350 B.C. Iron with a high carbon content is called *cast-iron* and is an easily melted, but brittle material, which can be cast into complicated shapes. It was produced in China as early as in 600 B.C.

Iron ore, which is found practically all over the earth's surface, is melted at high temperatures in furnaces and then alloyed with carbon to make steel. Modern structural steel is produced in two kinds of strengths: regular *structural steel,* that starts yielding under a stress if 36,000 pounds per square inch, and *high-strength* steel that yields at about 50,000 pounds per square inch. The first is used in most steel buildings, but the second has recently become almost as inexpensive as the first and is supplanting it, particularly in high-rise buildings. It is conceivable that high-strength steels may soon be developed, which will overcome the problem of the increase of brittleness with increasing strength. For example, one particular type of high-carbon steel, obtained by drawing steel through dies into thinner and thinner wires, is used to build cables and has already achieved ultimate strengths of 300,000 pounds per square inch. It can be used at stresses of up to 150,000 pounds per square inch. This means that we could safely hang the entire Leaning Tower of Pisa from a cable only 1.1 inches in diameter! Tiny pieces of steel, called *whiskers,* with a strength of one million pounds per square inch have also been produced. May we hope to go much beyond these very high strength values? Only so far. The attraction forces within steel crystals have a value of 4 million pounds per square inch. Nature puts a limit to steel strength, although chemists and physicists still have a long way to go before exhausting its capacity.

On the other hand, metallurgists have recently given steel new properties of great practical value. Although stainless steel does not rust, its production is too expensive to allow its use in the large amounts needed in construction. All steel structures, therefore, had to be painted at regular intervals (of a few years) to prevent rusting. It is only in the last ten years that a new type of *weathering steel,* labeled with proprietary names like COR-TEN by the U.S. Steel Company or Mayari R by the Bethlehem Steel Company, has solved this problem. Such steels have a rusting process which penetrates to a few hundredths of an inch from the surface and then stops, giving the steel a pleasant maroon color. The economies achieved by the elimination of continued painting have made this steel most popular for exposed structures and for the large pieces of sculpture called *environmental sculpture.*

Some steel properties present obstacles to its use in structures. Its 1,200°F. "melting" point and its minus 30°F. "transition temperature" to brittleness have already been mentioned. Another appears when steel more than two or three inches thick is used in the beams of very long bridges or the columns of very high buildings. Unless carefully treated thermally, these thick elements tend to slice into thin layers, like a piece of pastry, due to locked-in stresses. Only a few years ago, this dangerous phenomenon, called *lamination,* was the cause of the collapse of the West Gate Bridge in Australia and of others in other parts of the world. Luckily its causes are now understood and its dangers avoided. The tendency to move forward to greater achievement has often confronted technology with unsuspected dangers and motivated more thorough investigations of materials and structural behavior. The collapses of the West Gate Bridge and of the Tacoma Narrows Bridge have been blessings in disguise for the development of steel bridges.

Lamination stresses are not caused by apparent loads. Similar locked-in stresses may be caused in steel structures by the welded connections between elements. In welding, the steel of the two parts to be connected is melted at high temperature and a welding metal is deposited at the joint. When a well-executed joint cools, the connection becomes as strong as the steel of the jointed pieces. The high temperatures used in welding, if reached or cooled too rapidly and concentrated in too small an area around the joint, may produce thermal locked-in stresses, which the steel is unable to resist. They are similar to those in a steel hoop made by bending a straight strip into a circle and welding its two ends. Once the hoop is welded, it looks unloaded, but if one cuts it with a pair of shears, it snaps open, indicating that it had to be forced into its curved shape and, hence, was stressed. In order to avoid the locked-in stresses due to improper welding, welders are carefully trained and certified by law, like doctors. Moreover, welds are often inspected by means of x-ray and other refined techniques to detect faults in their execution.

Finally, it must be remembered that steel is "fatigued" by reversal of stress from tension to compression and vice versa, when this cycle is repeated many times. We use this phenomenon ourselves to break wire by bending it back and forth a number of times.

Thus there is more to steel design than loads and strength. The prudent engineer must be alert to a variety of dangers. As Thomas Alva Edison said to a young man he had to fire from his lab: "I don't mind the fact that you don't know much, yet. The trouble is that you don't even suspect." A good engineer should, of course, know, but should also be intelligently suspicious.

Reinforced Concrete

Possibly the most interesting man-made structural material is *reinforced concrete.* Combining the compressive strength of concrete and the tensile strength of steel, it can be poured into forms and given any shape suitable to the channeling of loads. It can be sculpted to the wishes of the architect rather than assembled in prefabricated shapes. It is economical, available almost everywhere, fire-resistant, and can be designed to be lightweight to reduce the dead load or to have a whole gamut of strengths to satisfy structural needs.

Concrete is a mixture of cement, sand, crushed stone or pebbles, and water. The water and cement paste fills the voids between the grains of sand and these fill the voids between the stones. After a few days the cement paste starts to harden or *set* and at the end of four weeks it gives concrete its nominal ultimate strength, which is as good as that of some of the strongest stones. Concrete mixtures are "designed" by specialized laboratories and mixed in strictly controlled proportions in concrete plants from which they are carried to the site in the revolving drums of large trucks, that keep mixing them en route. Concrete samples in the shape of cylinders or cubes are taken from each truckload and tested for compressive strength after seven and twenty-eight days. The strength of concrete depends on the ratio of water to cement, and of cement to sand and stone. The finer and harder the *aggregates* (sand and stone), the stronger the concrete. The greater the amount of water the weaker the concrete. The following episode illustrates in very human terms the importance of keeping the water-to-cement ratio low. During the construction of a terminal at J.F. Kennedy Airport in New York, the supervising engineer noticed that all the concrete reaching the site before noon showed good seven-day strength, but some of the concrete batches arriving shortly after noon did not measure up. Puzzled by this phenomenon, he investigated all its most plausible causes until he decided, in desperation, not only to be at the plant during the mixing, but also to follow the trucks as they went from the plant to the site. By doing so unobtrusively, he was able to catch a truck driver regularly stopping for beer and a sandwich at noon and, before entering the restaurant, hosing extra water into the drums so that the concrete would not start to harden before reaching the site. The prudent engineer must not only be cautious about material properties, but be aware, most of all, of human behavior.

Portland cement, as modern cement is called, is a mixture of limestone and clay, burned in a furnace and then pulverized. Impervious to water, it actually becomes stronger if submerged after it hardens. Samples of concrete taken thirty years after a concrete boat sank during World War I showed that the concrete had doubled its compressive strength. The Romans used a cement made only of lime to manufacture a concrete with aggregates of broken bricks and stones. This cement slowly dissolves in water, but it becomes almost as strong as modern concrete when mixed with *pozzolana,* a volcanic ash found at Pozzuoli near Naples. The Romans did not invent concrete, but a combination of pozzolanic concrete and

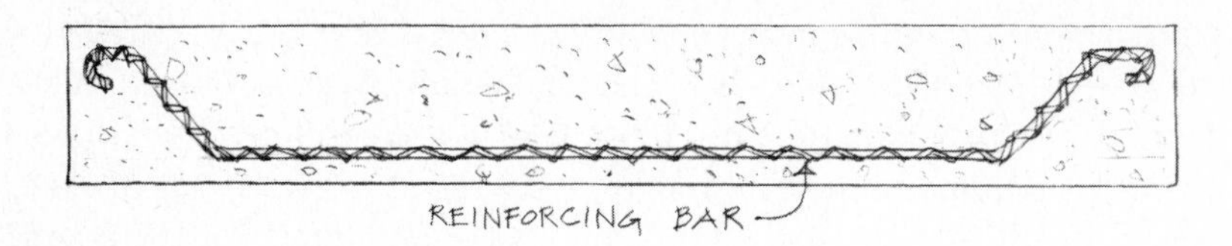

4.1 REINFORCED CONCRETE BEAM

outer surfaces of excellent stone, or good brick of burnt clay, allowed them to erect the majestic and massive structures which survive to this day.

Unfortunately, even the best concrete has a tensile strength barely one tenth of its compressive strength, a property it has in common with all stones. The invention of reinforced concrete remedied this deficiency and produced a structural material that, pound per pound, is the most economical. In reinforced concrete, bars of steel are embedded in the concrete in those areas where pulls will develop under loads, so that the steel takes the tension and concrete the compression. For example, the bottom of a beam supported at its ends is always in tension, while its top is in compression (see Chapter 5). Steel bars set near the bottom of the beam prevent the concrete from cracking under tension (Fig. 4.1) and make the beam work as if it were made of a material, like steel or wood, capable of resisting both kinds of stress. Reinforced concrete was originally invented in France towards the middle of the 1800s and its use spread very rapidly all over the world. It is today the most commonly used structural material.*

Ideal as it is for construction, concrete too has some unfortunate properties. If not properly wetted, or *cured,* while it hardens, it shrinks and cracks, allowing humidity to rust the reinforcing bars. Moreover it continues to stretch or shorten, *creeps,* under *constant* tension or compression loads, up to three or more years after hardening.

* The fact that the addition of a tensile material improves the performance of a compressive material was well known to the American Indians, and to the Babylonians and the Egyptians before them. Adobe is a paste of wet clay mixed with straw and sun-dried. The dried clay has a substantial compressive resistance and is stitched together by the tensile resistant straw. Some Berber cities in Morocco are circled by adobe defense walls, built centuries ago and standing to this day.

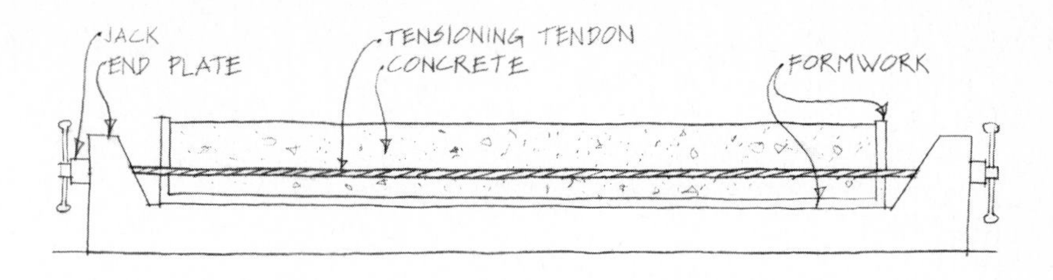

4.2 PRE-TENSIONED CONCRETE BEAM

Prestressing is the most recent procedure devised to improve the performance of reinforced concrete. The concrete at the bottom of a reinforced concrete beam tends to develop tiny (capillary) tension cracks due to stretching under the load of the steel bars it grips. If this concrete were compressed *before* the beam is loaded, the tension caused by the load would be counteracted by this initial compression and the beam would not develop capillary cracks. When the tension induced in the concrete by the loads is less than the initial prestressing compression, the concrete of the entire beam is always compressed and, as if it were steel or wood, is capable of resisting both types of stress from the loads. How is the precompression of the concrete obtained? In *pre-tensioning,* the steel bars are substituted by steel cables or *tendons,* made of exceptionally strong piano string wire, and are turn-buckled into tension against end steel plates (Fig. 4.2). Once the concrete hardens, the tendons are released from the plates and the concrete gets compressed by the tension in the steel tendons. The tendons's tension and the concrete compression balance each other and are locked-in stresses, invisible to an external observer.* In *post-tensioning* the tendons are threaded through plastic tubes set in the concrete, so that the concrete cannot grip them. After the concrete has hardened, the tendons are pulled against it and their tension puts the concrete in compression (Fig. 4.3). The post-tensioning operation can take place at the site after the beam has been erected, when the dead load is already acting on the beam, and can be adjusted as the loads on the beam increase.

The principle of prestressing is so simple that the reader may wonder why it was not put into practice soon after the invention of concrete.

* Pre-tensed concrete beams and columns are usually manufactured in special plants.

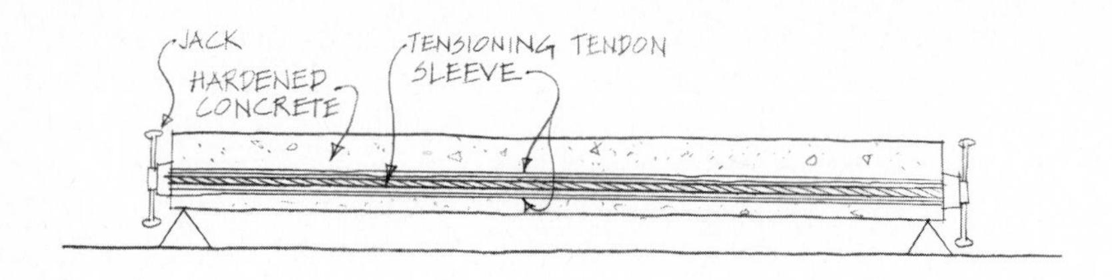

4.3 POST-TENSIONED CONCRETE BEAM

The French concrete engineers of the 1800s thought, indeed, of prestressing concrete, but they discovered to their dismay that the *creep* of the concrete in compression, together with the *yield* of the steel in tension, caused the loss of most of the prestressing action shortly after the beams were prestressed. As the compressed concrete crept and became shorter, the tendons slackened and should have been turn-buckled again to regain their tension. This was impossible once the structure was built. Prestressed concrete became practical only after high-strength steel became available at low prices. This steel can be pulled to such a high stress that even if some stress is lost through yield due to concrete shortening, it can still sufficiently compress the concrete. Once again, progress was due to an improvement in material properties rather than the inventiveness of the engineer. Today, prestressed concrete is widely used all over the world, because it is more economical than its cousin, reinforced concrete. Unfortunately, the high cost of the specialized labor required by prestressing has somewhat hampered its spread in the United States.

Plastics

The discovery, at the beginning of the twentieth century, of chemical methods of polymerization, that is, of producing compounds with long chains of molecules, has opened a Pandora's box to new materials. Under the name of *plastics* a great variety of organic materials, obtained by a combination of temperature and pressure, has been manufactured with an infinite variety of properties. Plastics can be made as strong as steel in both tension and compression, can be given an elastic or a plastic behavior, and are practically indestructible. Among the most useful plastics are those reinforced by glass fibers, like Fiberglas, which are shatterproof because glass, extremely strong in tension, has its brittleness cushioned by the plastic matrix in which it is embedded.

It may seem strange that plastics have not found important structural applications yet. This is due to two causes. In the first place, almost all plastics are more deformable than our present structural materials and, hence, cannot be used in the manufacture of large structural elements. Secondly, they are expensive. The economy of plastics presents a typical chicken-and-egg question: on one hand, they are expensive because they are used and produced in relatively small quantities (not large enough for the requirements of construction). On the other, they are produced in small quantities because they are expensive. Such a vicious circle is bound to be broken, and it is foreseeable that the role of plastics in structures will increase in importance as plastics better adapted to structural usage become less expensive.

For the time being, the most exciting structural application of plastics is that to high-strength fabrics in large inflatable roofs, dealt with in detail in Chapter 15. But years will have to pass before plastics can supplant materials as abundant, inexpensive, and well adapted to structural action as reinforced concrete, steel, masonry, and wood.

5 Beams and Columns

Newton's Laws

During the academic years 1665 and 1666, Isaac Newton, then in his early twenties, was prevented from attending Cambridge University by the spread of the plague throughout England. This was a blessing in disguise for him and a boon for humanity. Cloistered in his mother's home at Woolsthorpe, Newton was able to ponder the deep questions of physics he had set his mind to answer, and to develop the ideas of his gravitational theory. These ideas were so daring for the time as to approach absurdity. He postulated, for example, that bodies could exert forces at a distance without material contact, contrary to all physical evidence. Even so, his law of gravitation predicted exactly why the moon rotated about the earth in an elliptic trajectory in almost exactly 27 and 1/3 days and the earth around the sun in another elliptic trajectory in 365 days, 5 hours, 48 minutes and 46 seconds. It also "explained" why the legendary apple fell on his head. Not for 250 years would a genius of Newton's magnitude, Einstein, come up with another even more daring and abstract assumption, denying the existence of "action at a distance" and improving on the accuracy of Newton's results.

Newton, with the modesty of some great geniuses, never asserted to have explained why bodies attract each other. He simply stated that bodies behave *as if* they attracted each other with a force governed by his gravitational law. But, having at least assumed a cause of motion, he proceeded to describe how motion takes place. His three laws of motion allow the determination of the speed with which all bodies move, from the apple to the sun's planets. His first and third laws, when added to that of elasticity, are sufficient to solve almost all structural problems. His synthesis remains possibly the greatest in the history of science.

The first law of motion, as applied to structures, states that a body at rest will not move unless a new, *unbalanced* force is applied to it. The third states that, when a body is at rest, for each force applied to it there corresponds an equal and opposite balancing *reaction,* also applied to it. Since we want our structures *not* to move, except for the miniscule displacements due to their elasticity, Newton's laws of rest are the fundamental laws ruling the balance that must exist between all the forces applied to a structure.

In physics a body at rest is said to be in *equilibrium,* from the identical Latin word which means "equal weights" or balance. An understanding of two particularly simple aspects of the laws of equilibrium is essential to an insight into how structures work.

Translational Equilibrium

Consider an elevator hanging from its cables at a given floor. The pull of gravity on its cabin and occupants, that is, their total weight, acts downward. As the elevator does not move (is in equilibrium), a force upward must be exerted on it equal in magnitude to its total weight. This force can only be exerted by the cables. We conclude that the cables pull up on the elevator with such a force. The elevator is thus acted upon by two equal and opposite vertical forces and is in equilibrium in the *vertical* direction. According to Newton's third law, if we call action the force due to the weight, the pull of the cables constitutes the equal and opposite reaction. Similarly, if we stand on the floor, the pull of the earth (gravity) exerts a force pulling down on our body and, since we do not move down, an equal and opposite force must be exerted on it. This can only come from the floor pushing up on our feet, thus exerting a reaction (up) equal and opposite to the weight (down) of our body. Our body is in *vertical equilibrium.*

Consider now two groups of children pulling on the two ends of a rope. If neither group prevails, the pull of one group must be equal to that of the other and the rope, under the action of these two equal and opposite forces, is in equilibrium in the *horizontal* direction. If one group prevails, their force is greater than that of the opposing group and equilibrium is lost; the rope moves; it is not in equilibrium any more. In the first case, it may be hard to decide which of the two equal and opposite pulls is the action and which is the reaction. It is really a question of semantics and, perhaps, if one's boy or girl belongs to one group, on purely psychological grounds, we might be inclined to call this group's pull the action and the other's the reaction.

Identical considerations govern the equilibrium of structures. Since the total weight of one of the World Trade Center towers is approximately 140,000 tons and acts down, we know that the soil must exert on the foundation of the tower a force up of 140,000 tons. The soil under the tower must be quite hard to develop such a reaction. And so it is on Manhattan, an island which consists mostly of very solid rock. Similarly, if a wind exerts a horizontal pressure of, say, 30 pounds on each square foot of one of the tower's faces, which measure about 175,000 square feet, the total wind force on that face equals 30 times 175,000 square feet, or over 5,000 tons and the soil under the building must also react *horizontally*, in a direction opposite to that of the wind, with a force of 5,000 tons. If the soil were unable to do so, the tower would slide in the direction of the wind.

The two basic actions of tension and compression are usually developed in structural elements in equilibrium by equal and opposite forces acting along the center line or *axis* of the element. A steel column, of a weight usually negligible in comparison with the load it supports, is set in compression by the action down of the load's weight and the reaction up of its footing. One may say that the weight flows down the column until it is stopped by the footing. Similarly, the weight of the elevator pulls on its cables and we can visualize this pull travelling up along the cables until it is stopped at the top by the pulleys. One can "feel" axial tension or compression by pulling or pushing on a door-handle. The pull or push exerted by our body and balanced by the doorknob's reaction is felt by the muscles of our arm along which the pull or push travels.

From the Latin word *translare,* meaning to move in the same direction, equilibrium in a given direction is called *translational equilibrium.* For a body to be at rest it must move neither vertically nor horizontally. For example, a building will be at rest in total translational equilibrium if it does not move vertically or in either of the two directions parallel to its faces. Total translation equilibrium is thus satisfied by three conditions of translational equilibrium: vertical, parallel to one face, and parallel to the other face.

Rotational Equilibrium

If an empty cereal box is placed in front of an electric fan, it will move under the action of the air flow. If it rests on one of its long narrow sides, it will probably move horizontally, but if it rests on one of its short, narrow sides (its bottom), it will no doubt topple over before it moves horizontally. The cereal box is a good model for a tall building not well anchored into the ground and acted upon by a strong wind. Although in the second case the box was in vertical and horizontal equilibrium, since it did not slide or move up or down, it was not at rest since it toppled over. This motion occurs when the box turns around its lower edge on the side opposite to the wind or, as we say in physics, *rotates* around this leeward edge. To maintain the box at rest this rotational motion must also be prevented.

To understand the requirements of *rotational equilibrium* consider a see-saw with two children of identical weight sitting on opposite ends (Fig. 5.1a). The see-saw does not move vertically because the reaction up of its pivot balances the weights down of the two children (it is equal to twice the weight of one child). The see-saw does not rotate either, unless

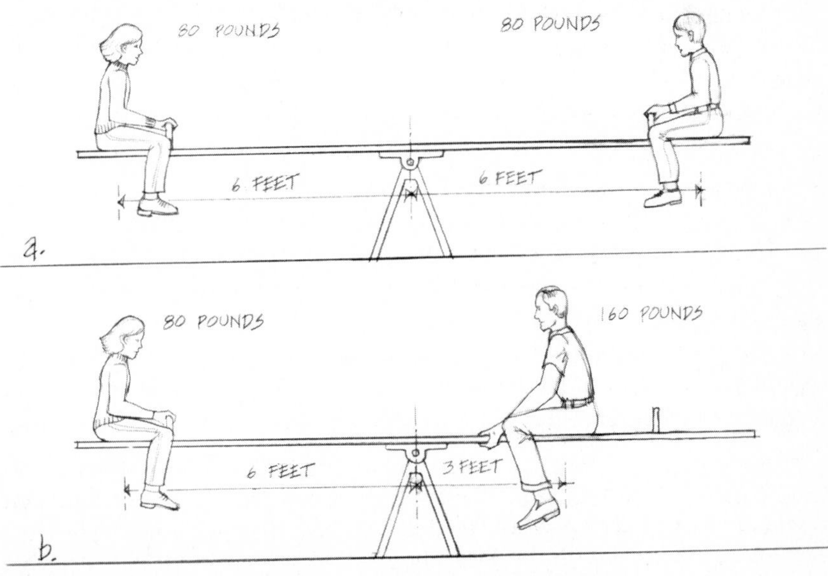

5.1 ROTATIONAL EQUILIBRIUM OF SEE-SAW

one of the children moves nearer the pivot, in which case the see-saw starts rotating down on the side of the other child and is not in rotational equilibrium anymore. Clearly rotational equilibrium requires not only that the weights of the two children be equal, but that their distances from the pivot also be equal. Let now a father weighing *twice* as much as his child sit on the see-saw and try to balance it (Fig. 5.1b). He succeeds if he sits *half* the distance of the child from the pivot. The two weights on opposite sides of the see-saw's pivot do not have to be equal for rotational equilibrium. What must be equal is the weight times its distance from the pivot. This is nothing else but the well-known law of the lever, which states that rotational equilibrium requires *equal products of forces times lever arms* (as distances from the pivot are called) tending to make the body rotate in opposite directions.

If one now fills the upright cereal box with enough sand or stones and blows on it again with a fan, the box will not topple over; it will be in rotational equilibrium. This occurs whenever the wind force times its vertical lever arm from the leeward edge is less than or at most equal to the weight of the building times its horizontal lever arm from the same edge (Fig. 5.2).* It is thus seen that the rotational action of a force can be balanced by that of another force in a different direction.

Simple as this may seem, the task of the structure is to guarantee translational and rotational equilibrium of the building under the action of any and all forces and reactions applied to it, including, of course, its own weight. The task of the engineer is to shape and dimension the chosen structural materials so that the structure may produce equilibrium without breaking up, and with acceptably small elastic displacements.

Beam Action

Equilibrium guarantees the stability of an entire building and, of course, of each one of its parts. In the California earthquake of 1971 a small hospital building toppled over without suffering serious damage. Its strength was sufficient, but not its equilibrium capacity. To prevent the toppling of a very high building, when the building's weight is not sufficient to counteract the wind, its columns are anchored into a founda-

* For example, in one of the World Trade Center towers, the horizontal wind force of 5,000 tons has a vertical lever arm of 700 feet (half the tower height) and a product equal to $5{,}000 \times 700 = 3{,}500{,}000$ or 3.5 million ton-feet. The tower weight of about 140,000 tons has a horizontal lever arm equal to 60 feet (half the tower width) and a product equal to $140{,}000 \times 60 = 8{,}400{,}000$ or 8.4 million ton-feet. The weight wins and the tower does not topple over.

tion deep in the ground. It this case the building bends slightly under wind pressure, but remains in rotational equilibrium in part due to the counteraction of its weight but also due to the forces exerted by the anchored columns. The windward columns are set in tension and those on the leeward side in compression (Fig. 5.3), thus creating a tendency to turn the building against the wind. The wind drift or lateral displacement of the top of the building is due almost entirely to the lengthening of the windward columns under tension and the shortening of the leeward columns under compression. Since the floors are rigidly connected to the columns, they tilt slightly and remain at right angles to them. All these deformations are so minute that they cannot be detected by the naked eye, but give to the building a slightly deflected curved shape.

The shortening of a column under the action of an allowable compressive axial force or the lengthening of a cable under an allowable tensile force is extremely small. In the building under lateral wind load, we find, however, that the *lateral* deflections may be substantial. The columns of a 1,000-foot skyscraper shorten under the vertical loads by only one tenth of an inch, but the top of the skyscraper may bend out several feet under the action of lateral loads. The deflections due to loads perpendicular to

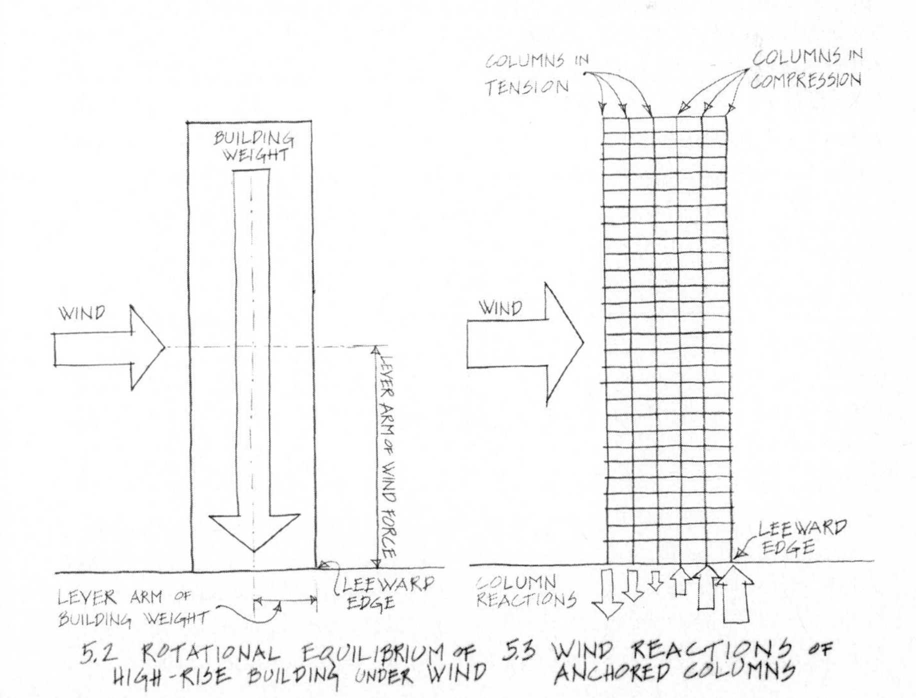

5.2 ROTATIONAL EQUILIBRIUM OF HIGH-RISE BUILDING UNDER WIND

5.3 WIND REACTIONS OF ANCHORED COLUMNS

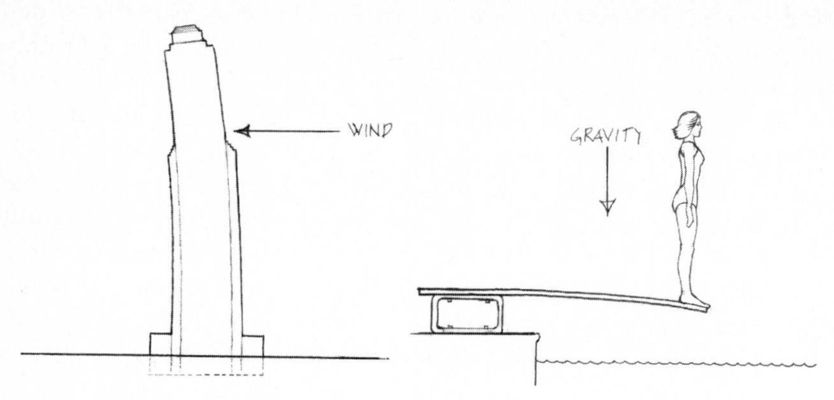

5.4 BUILDING BENT BY WIND AND CANTILEVER BENT BY LOAD

the structure are much larger than those due to axial, longitudinal loads. They are called *bending deflections* and are typical of basic structural elements, called *beams,* which are usually loaded at right angles to their longitudinal axis. A skyscraper under wind acts like a gigantic vertical beam stuck into the ground, much as a diving board acts in supporting the load of a diver (Fig. 5.4). Let us look in greater detail into the deformation of these *cantilevered beams.*

Consider a beam stuck into a wall and carrying a load at its tip—for example, a wooden jumping board with a man on the end. Under the man's weight, the beam deflects and its tip displaces downward. The beam bends, because the fibers of its upper part become longer and those of its lower part shorter. If one draws vertical lines on one side of the beam, these remain straight and perpendicular to the beam's upper and lower surfaces (Fig. 5.5). The lengthening and shortening of the upper and lower wood fibers are shown by the crowding of the vertical lines at the bottom and the opening-up of the lines at the top. The most remarkable feature of this deformation is that the fibers midway between the top and bottom maintain their original length. They are neither stretched nor shortened. Hence, they are neither under tension nor compression; they are unstressed. Moreover, we may also notice that the lengthening and shortening of the fibers in the upper and lower halves of the beam increase from being nonexistent at the middle fiber to the greatest values at the top and bottom of the beam, respectively. Since the board's behavior is linearly elastic, the tension and compression in the beam fibers grow linearly from zero at the middle to maximum values at the

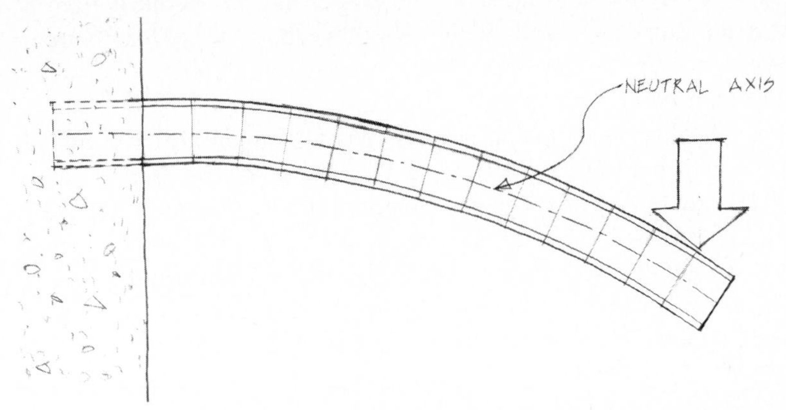

5.5 SECTIONS AND NEUTRAL AXIS IN BENT CANTILEVER

top and bottom fibers. This explains why the short lateral lines remain straight, as Leonardo first and then Navier knew (see Chapter 1).

In the compression of a column or the tension of a cable the loads are evenly divided between all the fibers of these elements. In a beam, instead, the extreme lower and upper fibers are highly stressed, while all the other fibers are stressed less and less as we approach the middle fibers. These middle fibers do not do any work while the others do more and more work the nearer they are to the top and bottom of the beam. Thus, most of the beam material is not utilized to its maximum capacity. A beam, whatever its material, is not a very efficient structural element as, on the average, its fibers work at half its allowable capacity.

In a beam supported at its ends on columns, the crowding towards the top of the vertical lines on its side and their opening-up towards the bottom (Fig. 5.6) indicate, again, that the lower fibers are in tension and

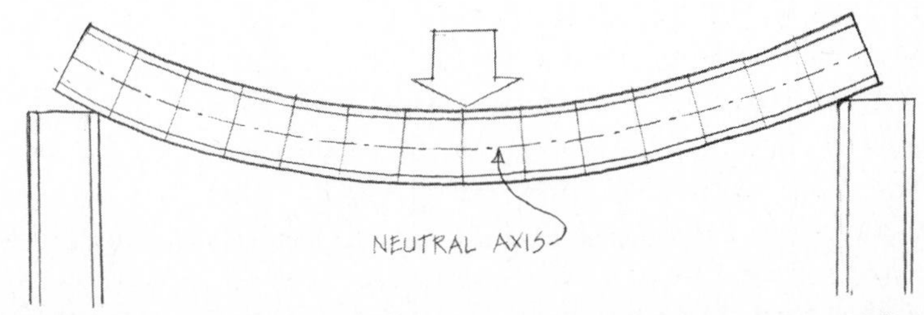

5.6 SECTIONS AND NEUTRAL AXIS IN END-SUPPORTED BEAM

the upper in compression. This is why cracks appear at the bottom of beams made out of materials with a low tensile capacity, like stone or concrete, and why steel reinforcing bars are set at the bottom of concrete beams supported at their ends.

In channeling loads to the ground our most important and difficult task is to span horizontal distances so that loads may be carried across them. The structure of a bridge has no other purpose. Similarly, the floors of a building allow people to walk over the heads of people on lower floors. It is easy enough to support a load with a column right under it; it is much more difficult to carry the weight of a truck to the bridge supports so that the river may flow unimpeded below it. In buildings, distances (of up to 100 feet) are spanned by numerous horizontal beams. It is unfortunate that beams should be rather inefficient.

Is there anything one can do to improve the efficiency of beams? Realizing that all the material in the neighborhood of the middle fiber, which is called the *neutral axis,* is understressed, the thought occurs to move this material away from the neutral axis toward the top and bottom of the beam. If this is done as shown in Figure 5.7, the shape of the beam's transverse cut, or *cross-section,* becomes similar to that of a capital I. Of course, one cannot displace all the material near the middle fiber up to the top and down to the bottom of the beam. Some material must always connect the top and bottom parts of the I-beam, called its *flanges,* or they would become two separate, thin and flexible beams. This narrow vertical strip is called the beam's *web* and is typical of the steel beams used in the construction of high-rise steel buildings.

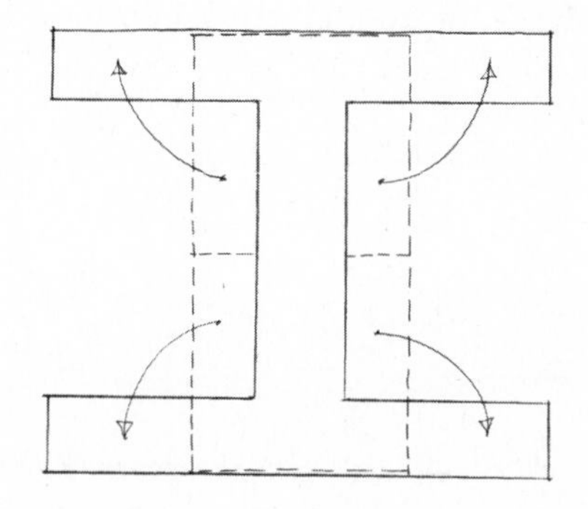

5.7 I-BEAM WITH SAME CROSS-SECTION AREA AS RECTANGULAR BEAM

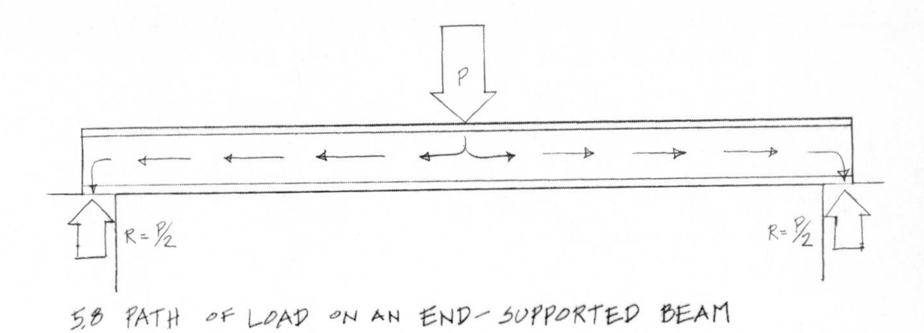

5.8 PATH OF LOAD ON AN END-SUPPORTED BEAM

I-beams of steel with wide flanges, called *wide flange sections,* are obtained by rolling heated and softened pieces of steel between the jaws of powerful presses and have flanges much wider than the top and bottom segments of a capital I. This is the most efficient shape a beam can be given to carry vertical loads horizontally from one point to another. One may think of a beam as a structural element that transfers vertical loads to the end supports along its horizontal fibers, as if the beam deflected the vertical flow of the loads by ninety degrees only to turn them around again in a vertical direction at the beam supports (Fig. 5.8).

Beams are made out of steel, aluminum, reinforced concrete, and wood. Their downward-bending displacements under load, larger than those of columns, must be limited lest they dip impossibly or the plaster begin to flake off the ceiling. Codes limit the bending deflection of a beam to less than the beam length divided by 360. The stiffness of a beam shape is increased by shifting part of the material away from its middle fibers and is measured by a quantity called the *moment of inertia* of the beam's cross-section, given in all beam manuals. Thus, deep beams are stiffer than shallow beams. On the other hand, the beam stiffness diminishes dramatically with increases in length; doubling the length of a beam makes it sixteen times more flexible.

If tension is characterized by lengthening and compression by shortening, *beam action* or *bending* is characterized by the curving of the beam. Whenever a straight element becomes curved under load, it develops beam action and the more curved it becomes, the larger its bending stresses. When a beam's ends curve up, as in a beam supported at its ends, its lower fibers are in tension and its upper fibers in compression. Whenever a beam curves down, like a cantilever supporting a balcony, the upper fibers are in tension and the lower in compression. (Hence, in

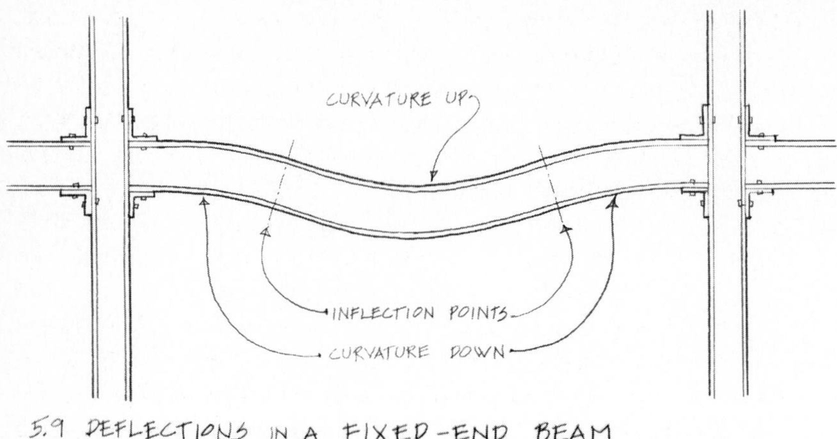

5.9 DEFLECTIONS IN A FIXED-END BEAM

a balcony beam of reinforced concrete, the reinforcing bars must be located towards the tensed top of the beam). In general, then, the tension appears on the opposite side of the curvature (see Figs. 5.5 and 5.6).

It has been tacitly assumed so far that a beam supported at the ends on two walls or two columns is not rigidly connected to the supports so that its ends rotate due to the bending deflections (see Fig. 5.6). Most of the time, instead, the beam is *fixed* into the supporting walls or rigidly connected to the top of the columns. In a reinforced concrete beam the simultaneous pouring of the concrete makes the columns monolithic with the beams, while in a steel structure beams are rigidly bolted to the columns. Such beams have ends which cannot rotate (or rotate only slightly). They are called *fixed-end* or *built-in* beams.

Fixed-end beams are particularly stiff and strong. They carry one-and-a-half times the load of a supported beam of the same length and they deflect five times less. Fixed-end beams deflect under load as shown in Figure 5.9. The figure shows that rigid connection to the supports makes the beam ends curve down, but that its middle curves up. This is why in the neighborhood of the supports of a fixed-end concrete beam the reinforcement is set near its top and in its middle portion near its bottom. There are two points in the beam at which the curvature changes from down to up. For a short length near these points the beam has no curvature and hence, develops no bending stresses. Such points are called *points of inflection*. Thus, the deflected beam shape allows a visualization of the distribution of bending stresses in the beam.

Lest it may be construed from this brief discussion of bending that the reinforcing bars in a concrete beam should always appear either near its top or its bottom, let us notice that a vertical column stuck into its foundation acts like a vertical cantilever under lateral loads and may be hit by the wind from either side. When the wind blows from the right, its fibers in tension are on the right and its compressed fibers on the left, but when the wind blows from the left, the role of the fibers is interchanged. In such columns reinforcing steel must appear near both sides of the column. In addition, the reinforcing bars serve to "stitch" the concrete and keep it together. To this purpose in most beams the bars, set at top and bottom, are connected by vertical hoops of steel, called *stirrups*, constituting a steel cage containing the concrete (see Fig. 5.12). The concrete, in turn, covers the steel bars, thereby preventing them from rusting. An insufficient concrete cover of the bars, which allows water to penetrate the beam and rust its reinforcement, has been known to make the reinforcement crumble inside the concrete until the structure collapses. This is a dangerous phenomenon because the rusting of the bars may not show on the outside of the structure to give warning of its condition.

Shear

When a cantilever is loaded at its tip, it tends to rotate, but its rotational equilibrium is guaranteed by the action of the tensile forces in its upper fibers and the compressive forces in its lower fibers, which tends to make the beam rotate in the opposite direction (Fig. 5.10). Obviously its translational equilibrium must also be satisfied. This means that since the tip load acts down on it, an equal and opposite force or reaction must act up on the cantilever. This can only happen where

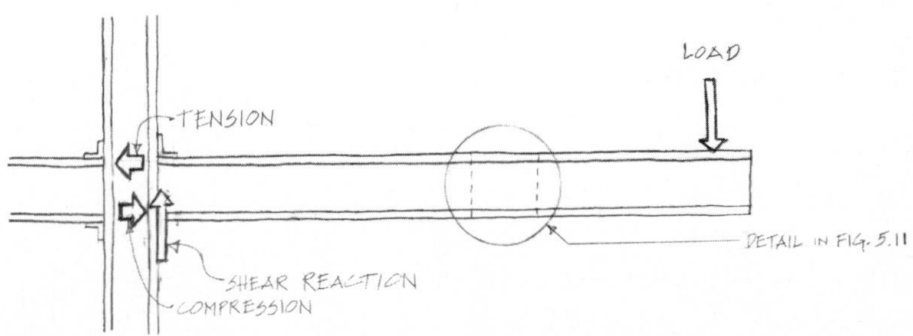

5.10 ROTATIONAL AND VERTICAL EQUILIBRIUM OF CANTILEVER

the beam is supported. Hence, the support must exert an upward reaction equal to the load (Fig. 5.10), which is called *shear* because, together with the tip load, it exerts on the beam the type of action that shears exert on a steel sheet while cutting it.

Tension tends to move the particles of the material apart; compression pushes them together; shear makes them slide one with respect to the other. It may be rightly thought that shear action is a new and different type of structural action and that there are three *elementary* structural actions rather than two, as stated in Chapter 4. This is not so. That shear action is a combination of tension and compression is shown in Figures 5.11 a,b,c. Figure 5.11a shows a small cube cut out of a cantilever beam with two equal vertical and opposite forces acting on its left and right faces, the load down and the shear reaction up, which together guarantee the vertical equilibrium of the cube. But their opposite directions together with the distance between them tend to make the cube rotate clockwise, just as pulling with our right arm and pushing with our left tends to turn the wheel of our car clockwise. To guarantee the rotational equilibrium of the cube, equal and opposite forces, or shears, tending to rotate the cube counterclockwise *must* appear on its horizontal faces as shown in Figure 5.11b. This indicates that whenever shear tends to slide the particles of a beam vertically with respect to one another, it also necessarily tends to slide them horizontally. The equivalence of shear to a combination of tension and compression can now be seen in Figure 5.11c. This figure shows how the upper and left shears actually combine to become a tensile force directed up and to the left, while the lower and right shears combine to become an equal tensile force acting down and to the right. But the upper and right shears also combine to become a compression acting down and to the left, while the lower and left shears combine to become an equal compression up and to the right. It is thus

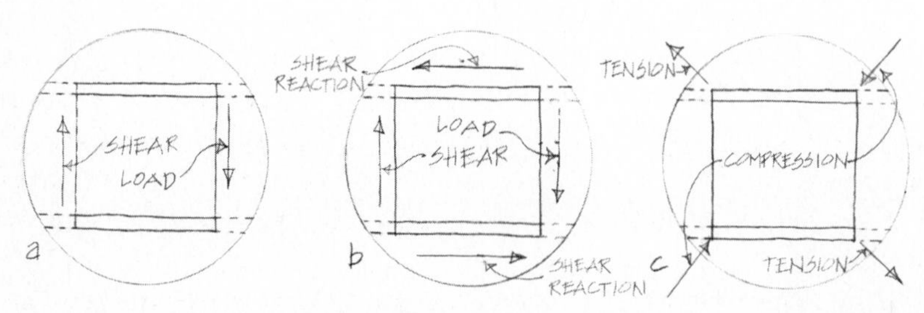

5.11 EQUIVALENCE OF SHEAR TO TENSION AND COMPRESSION AT RIGHT ANGLES

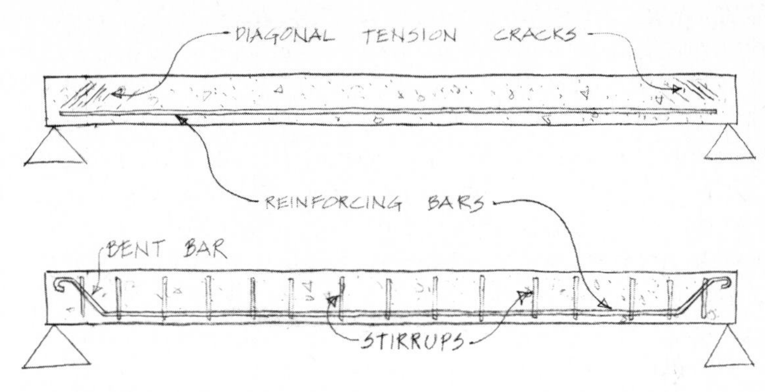

5.12 SHEAR TENSION-CRACKS AT BEAM SUPPORT

seen that shear is structurally equivalent to tension and compression at right angles to each other and at forty-five degrees to the shears. A physical proof of this equivalence is given by a reinforced concrete beam insufficiently reinforced against shear. Diagonal cracks at forty-five degrees appear near its support showing that the concrete tensile strength has been overcome by the tension component of the shear. Such cracks are avoided by bending the lower bars of the beam at an angle of forty-five degrees near the supports so as to absorb the tension due to shear, or by "stitching" the horizontal concrete layers with stirrups to prevent their sliding with respect to one another (Fig. 5.12).

Almost none of the problems presented by concrete beams exist in steel beams, since steel has as good a resistance to tension as to compression. And yet the next section will show that trouble can arise even in steel beams.

Buckling

It may seem strange that in a modern steel building the columns should have the same wide-flange shape as the beams, when this I-shape was shown to be ideal for bending while columns are not bent, but compressed. But this choice eliminates one of the most dangerous structural phenomena called *buckling:* the bending of a straight element under compression.

If one pushes down with increasing pressure on a vertical thin steel ruler supported on a table (Fig. 5.13), the ruler at first remains straight,

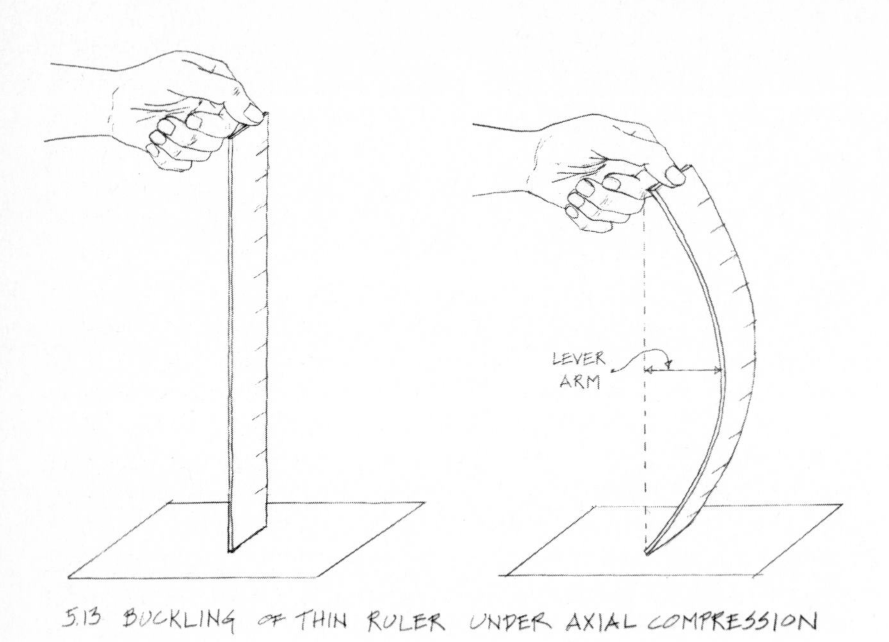

5.13 BUCKLING OF THIN RULER UNDER AXIAL COMPRESSION

but there comes a point when, rather suddenly, the ruler bends out. It can be proved that a *perfectly* straight ruler acted on by a compressive force *perfectly* aligned with its axis will bend out suddenly at a given value of this force, called its *critical value.* (The real ruler does not bend so suddenly because it is never perfectly straight.) The moment the ruler bends out, the compressive force acquires a lever arm with respect to its axis (Fig. 5.13) and bends it progressively more. This is a chain reaction where the more the ruler bends, the larger the lever arm becomes. This increases the bending action of the force, which increases the lever arm, and so on. Very soon the ruler fails in *bending*. The column is said to become *unstable* when the load reaches its critical value.

This behavior is typical not only of thin columns, but of any thin element under compression and has acquired great significance due to the thin sections allowed by modern, strong materials. The columns of a Greek temple could never buckle because they were chunky and short. The slender columns of a modern building are much more likely to buckle. Since buckling is a phenomenon involving bending, it becomes clear why modern steel columns have the shape of wide-flange beams.

Their resistance to buckling is magnified by this shape without a costly increase in material.

The load capable of buckling a column, or its *critical load,* depends on the slenderness, the material, and the way a column is supported. Since the moment of inertia was found to be a quantity characterizing the amount of material moved away from the neutral axis of a beam and hence measuring its bending stiffness, it is not surprising that the buckling load increases in proportion with the moment of inertia of the wide-flange columns. The longer a column, the slenderer it becomes and its buckling load is reduced in proportion to the square of its length. A column twice as long as another has a buckling load four times smaller. A column stuck into its foundation and free to move at its top (a cantilevered column) has a buckling load eight times smaller than the same column stuck into its foundation and rigidly connected to a floor at its top (Fig. 5.14). Finally, the stiffer the column's material, the stronger the column. A steel column is three times stronger against buckling than an identical aluminum column.

It is interesting to notice that buckling is a consequence of the basic "least work law" of nature. If an increasing load is applied to the top of

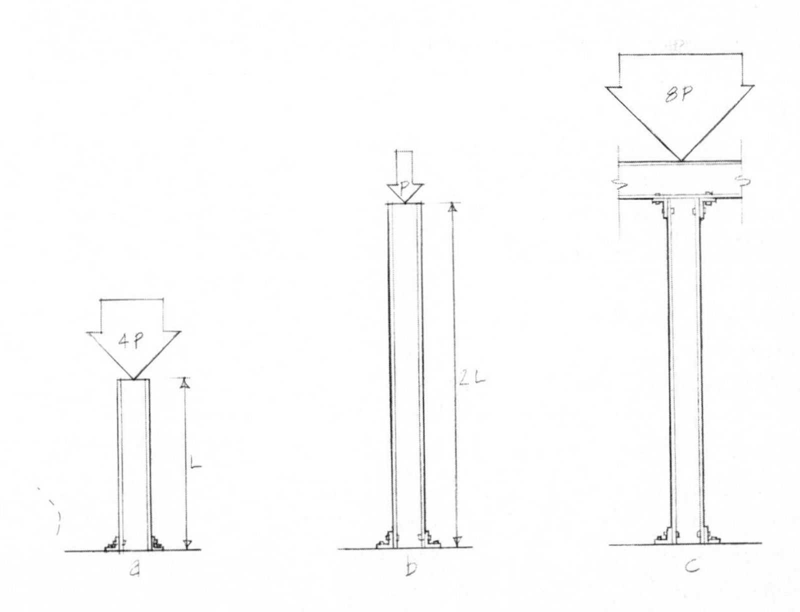

5.14 BUCKLING LOADS OF (a AND b) CANTILEVERED AND (c) FIXED-END COLUMNS

a column, the load first comes down by compressing and shortening the column. Since loads always tend to settle in their lowest balanced position, this type of lowering of the load would go on indefinitely, were it not for the fact that at a certain value of the load, the column can lower it further in two ways: either by continuing to shorten in compression or by bending out. As soon as the work required to further lower the load is less in bending than in compression, the column follows the easier path and bends. The column buckles.

The shape given modern steel columns may prevent the buckling of the column as a whole, but may still allow the buckling of some of its parts when these are thin. Thin webs are most sensitive to local buckling and, at times, must be stiffened to prevent it. Buckling of the web of steel beams may occur near the supports due to the compressive component of the shear. The lower flange of a cantilevered wide-flange beam, being thin and compressed in bending by vertical loads, may also buckle. This buckling occurs in a lateral direction, as shown in Figure 5.15, and twists the beam, besides bending it.

Buckling is one of the main causes of structural failure. The roof of the Hartford Civic Center hockey rink was a *space frame* (described in Chapter 9) made out of steel bars. It covered an area of 360 feet by 300 feet and was supported on four massive pillars of concrete set 40 feet in from its corners. Its upper bars were compressed by its enormous dead load of 1,400 tons and by the weight of snow, ice, and water accumulated on it. After standing for four years, it suddenly failed, at four o'clock in the morning following a heavy snowstorm in 1978. Four-

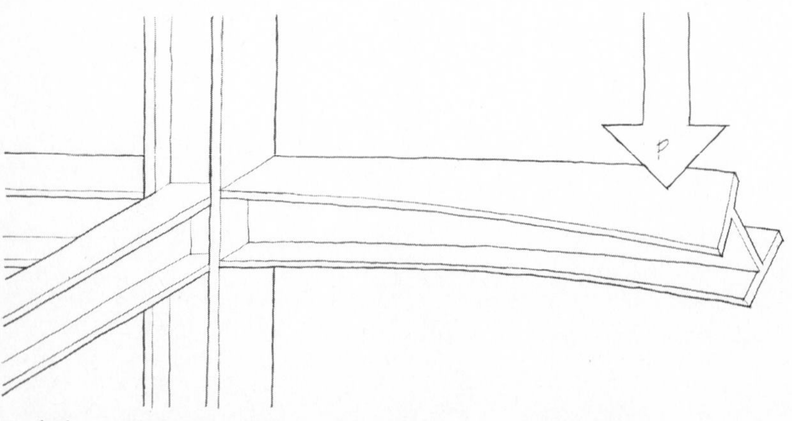

5.15 LATERAL BUCKLING OF WIDE-FLANGE BEAM

teen hundred tons of steel came crashing down on the floor and stands of the hockey rink, where only five hours earlier 5,000 spectators sat watching a game. The roof collapsed in less than ten seconds. This structural failure is attributed by engineers entrusted by the city of Hartford with its investigation to the buckling, at first, of only a few compressed bars, which shifted to adjoining bars the load they were supposed to support. The overloaded adjoining bars in turn buckled and the progressive spreading of buckling to more and more bars produced the collapse of the entire roof. This dramatic occurrence shows that one of the most dangerous characteristics of a buckling failure is its suddenness, which gives no warning. Whenever a structure under load chooses the easy path of bending rather than the foreseen path of compression, the structure may fail. Good engineering judgment, correctly shaped and supported elements, strong materials, and careful supervision during construction are needed to avoid this particularly tricky and sensitive structural behavior.

6 Houses

Prehistoric and Historic Houses

The housing of primitives was usually, but not always dictated by the environmental climate and the availability of structural materials. The use of wood prevailed wherever this almost ideal material abounded, but its natural decay, due to both humidity and fire, has left few, if any remains of its use by primitive peoples. Even so, we can infer from the housing of the primitive tribes still living in our time that huts were first erected by leaning tree trunks or branches one against the other in a circular pattern so as to obtain a conical structure, which was then covered with mud. As centuries and millennia went by, the basic cell of the house acquired a rectangular plan. Walls of vertical trunks and mud appeared in cold climates together with roofs of wooden beams running from wall to wall made weatherproof by means of mud or earth sods. In hot climates walls were made of small branches or cane, and roofs were thatched. The dimensions of the cell were limited by the length of the available timber, until interior columns were introduced to reduce the beams's spans or to allow the use of long beams which would have been too flexible without intermediate support (Fig. 6.1). Finally, spans between walls were made longer, by use of main beams in the shorter direction of the rectangular plan and of secondary beams spanning the main beams. The essentials of the structural system to be used in houses for thousands of years had been established.

This structural system delegated the resistance against the wind and other horizontal forces to the outer cantilevered walls. The minor resistance of the walls to bending and the weak connections between columns and beams, tied by vegetable ropes, made them easily damaged or destroyed by winds and earthquakes. This liability was not overcome until the Middle Ages, when solid, nailed joints between wooden elements were invented, and modern times, when monolithic construction was made possible by mortar, concrete, and steel.

The solution to the housing problem took a totally different form in those areas of the world, like Asia Minor, Greece, Crete, Sardinia, southern France and England, where stone was abundant and wood scarce. In a band running uninterruptedly through this area prehistoric houses dating back as early as 5000 B.C. still stand, characterized by a round plan, stone walls, and stone roofs. Houses with as many as three interconnected circular rooms and two stories are found among the *nuraghe* of Sardinia. One of these has even a spiral outer staircase connecting two superimposed floors. The "domed" roofs of this type of house, the *corbelled domes*, are obtained by circular layers of stone slabs, each cantilevering inward from the layer supporting it and wedged in by the layer above it (Fig. 6.2). The trulli (see Fig. 1.2), the traditional housing of the villages in a large area of the Puglie region of Southern Italy, are identical to the nuraghe except for the whitewash inside and out.

6.1 ROOF BEAMS SUPPORTED BY INTERIOR COLUMNS

6.2 CORBELLED STONE DOME

The palaces and houses of the Minoan civilization on Crete, which were built between 3000 and 2000 B.C., have rectangular plans with outer stone walls and intermediate stone columns. They apparently were covered by wooden-structured roofs, which have disappeared. The palaces had plastered walls covered with the most brilliant frescoes (some have survived) and large windows and doors framed by wooden beams. The bending resistance of wood allowed these large openings in stone walls. The wooden *lintels* at the top of the openings, supporting the stone blocks above them, were propped by wooden columns on the side of the openings. Airy and light housing with as many as four floors was thus created, ideally suited to the warm, at times tropical, climate of Crete. The Minoan civilization, which suddenly disappeared toward the middle of the sixteenth century B.C., was transplanted to Greece, giving rise to the Mycenaean cities, again built with stone walls and wooden roofs or corbelled vaults.

In extreme climates solutions of a different character were adopted. In the jungles of Amazonia in South America some tribes lived in extended households with as many as one hundred people. Their houses consisted of vertical wooden posts set in a large circular pattern, connected at their tops by peripheral wooden tree branches and secured by radial branches to vertical posts set in a smaller, inner, concentric circle (Fig. 6.3). A platform, raised a few feet for protection from the poisonous animals creeping on the ground, and a thatched roof completed the house. There were no walls and the cool wind could flow through unimpeded. What made this scheme interesting was the procedure used to protect privacy in the absence of partitions of any kind. The method was as simple as it was effective: if the occupants of the house were looking towards the outside, they were "in public" and behaved accordingly. But, if they looked toward the inner posts, they became "invisible" and could perform the most intimate acts of life unnoticed by their large family.

In the arctic the Eskimos invented the igloo, a dome in the shape of a perfect half-sphere made out of snow blocks, which by its form is aerodynamically efficient in reducing the wind pressure on the building and at the same time produces one of the strongest structural shapes devised by man. (The strength of a dome is superior to that of almost any other structural form, as we shall see in Chapter 13). The fire set in the middle of the igloo melted a thin layer of snow, that became ice when the fire was extinguished, making the igloo totally impermeable to the wind. Made of a material with unlimited availability, the igloo could have been a permanent abode, but was easily erected in a few hours when the Eskimos moved in search of food. Their mores allowed them to abandon their elders to freezing death on the ice pack when

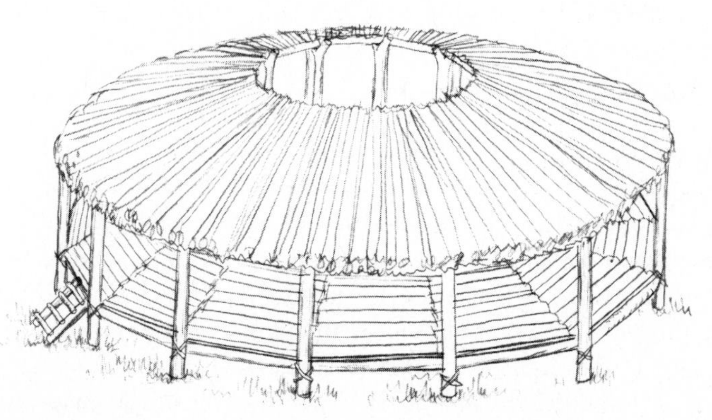

6.3 AMAZONIAN HOUSE

they became unable to undertake the long marches from one location to the other. On the other hand Eskimo hospitality insisted that a solitary male traveller sleep in the same sleeping bag with the host's wife. It seems that concepts of modesty and morality are heavily influenced by the climate and the housing it demands.

The use of sun-dried clay and adobe, a material relatively strong in tension and compression, was common in the dry countries of the East, and the later invention of fired clay bricks made possible the development of the Roman house, which was to become typical of the houses of the Middle Ages. These had not only brick walls, but roofs of clay tile supported on wooden inclined beams and superimposed horizontal beams, called *purlins*, a tradition which has survived to this day in many parts of France, Italy and other countries throughout the world. Walls of fired clay brick, held together by mortar (a mixture of lime, sand, and water), have a certain amount of resistance to tension, and thus to bending, and a remarkable resistance to compression. The tower of the City Hall of Siena in Italy, known as the Torre del Mangia (from the nickname of one of its bellringers, famous for his gargantuan appetite), is built entirely of mortared brick and reaches a height of 334 feet. Although built in 1348, it is even today the tallest brick tower in the world. The height of the masonry lighthouse at Pharos in Alexandria, Egypt, erected in 300 B.C., which collapsed in A.D. 1326, is said by some historians to have been higher, either 370 feet or even 600 feet.

Houses with *bearing walls* of brick and mortar and with floor beams of wood could be many stories high due to the partial bending resistance of the walls against wind. The use of steel beams made possible longer spans and modern housing with bearing walls of brick (stiffened by light reinforced concrete columns and beams) reaches today up to twenty-seven stories.

The American House

In the United States, as anywhere else, traditional houses were built of the most commonly available materials. In the Southwest, the mixture of clay and straw called *adobe* was used to build houses with thick walls, warm in winter and cool in summer. In the Southwest the octagonal *hogan* of tree trunks and mud was used by many Indian tribes. In the Great Plains the pioneers used the sods cut by the plow in regular blocks to build walls and sometimes roofs of houses that were safe against fire, wind and earthquake. In wooded regions of the north the log cabin, easy to build, with walls and roof tightly joined, safe against wind and comfortable in

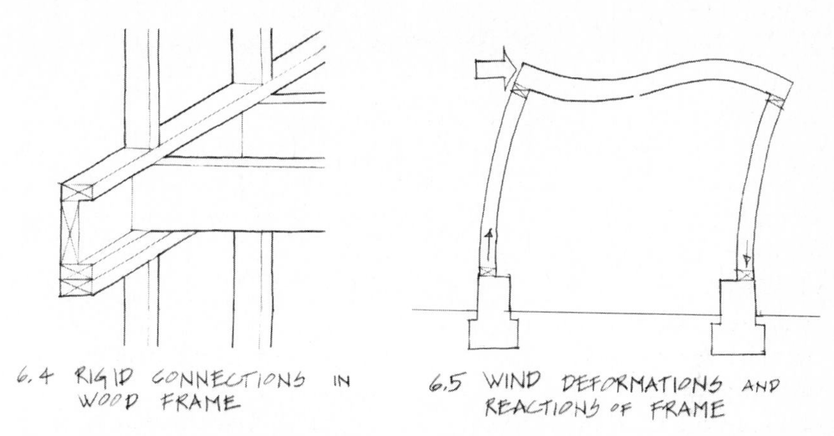

6.4 RIGID CONNECTIONS IN WOOD FRAME

6.5 WIND DEFORMATIONS AND REACTIONS OF FRAME

with walls and roof tightly joined, safe against wind and comfortable in all seasons, has remained so popular that to this day two million of them are built a year, all the way from do-it-yourself models to mansions costing hundreds of thousands of dollars.

The abundance of wood, new treatments against fire and insects, and standardization of structural elements in the shape of beams and columns of rectangular cross-section—the so-called 2 by 4s, 2 by 6s or 4 by 10s—have made the one-family wooden house our most popular, even today. Limited to one, two, or at most three or four stories, these houses are supported by wooden columns, rigidly connected to the floor and to the roof beams. These rigid connections make them strong against lateral forces, because any force trying to bend the columns must also bend its beams and make them help carry the loads (Figs. 6.4, 6.5). Columns and beams are said to act as a *frame*. When one side of the frame is hit by the force of the wind, its beam transmits half of it to the other side by compression, while by virtue of the rigid connections the turning action of the wind is mostly resisted through the upward reaction of the leeward column and the downward reaction of the windward column (Fig. 6.5). The floors of narrowly spaced *joists* with nailed plywood covering increase the horizontal stiffness of the house besides carrying the floor loads (Fig. 6.6).

Thus, an all wooden house with a structure of frames and triangular roof trusses, and with walls of sheathing and shingles, is capable of resisting the usual vertical loads of snow and the heavy forces due to the wind, except, at times, those of hurricanes and tornadoes. Since such houses can be built with simple tools, carpenters have always been responsible for a

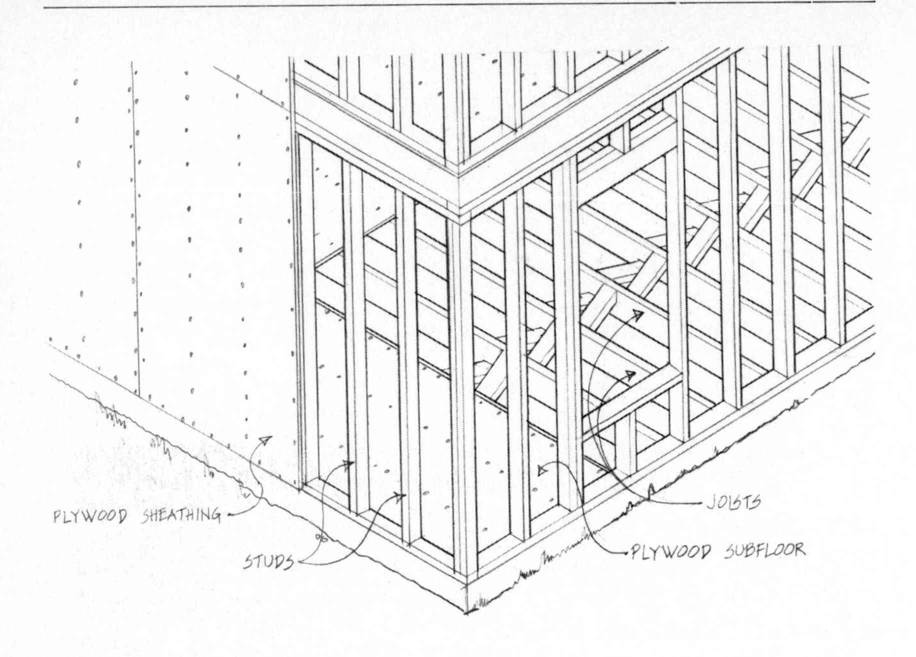

6.6 WOOD FRAMED CONSTRUCTION

large percentage of the housing built in the United States. Framed and trussed wood buildings have been erected in Europe and other parts of the world since the Middle Ages, and houses similar to the American one-family house are built today in all countries having large availability of timber. The precutting of wood beams and columns to standard sizes, the manufacture of inexpensive mechanized tools, and the production of easily installed heating and plumbing systems have perfected the old structural and mechanical systems, making the one-family house widely available. Entire precut houses are sold to be assembled by the owner himself. These and the precut log-cabin houses are examples of prefabrication.

Concrete Frame Housing

Frame action against wind and earthquake forces requiring rigid connections between the columns and the beams of the frame is inherent in reinforcing concrete construction. Framed structures of reinforced concrete were built from the very inception of this modern material and are used all over the world, mostly for housing purposes.

In a reinforced concrete building, heavily reinforced concrete foundation blocks, called *footings*, are first poured in the ground, often below its original level or *grade* in order to better utilize the available site area through the construction of one or more underground floors or basements. This is of the utmost importance in large cities like New York where land may cost sixty or more dollars per square foot or nine or more million dollars for a high-rise building plot of 120 feet by 120 feet.

Some of the footings' reinforcement is left protruding and hollow vertical wooden or steel forms for the columns are built around them over the footings. A cage of vertical reinforcing bars and horizontal thinner *stirrups* (Fig. 6.7) is first lowered into these forms and concrete is then poured into them, leaving a length of vertical bars to stick up from the forms. Once the column concrete has set, a horizontal surface of wood planks, propped at many points by wooden columns or *struts*, is erected at the level of the column tops. Reinforcing bars for the floor are laid on top of the planks in a rectangular grid, supporting them on small blocks called *chairs* one or two inches above the planks. The bars of the floor are

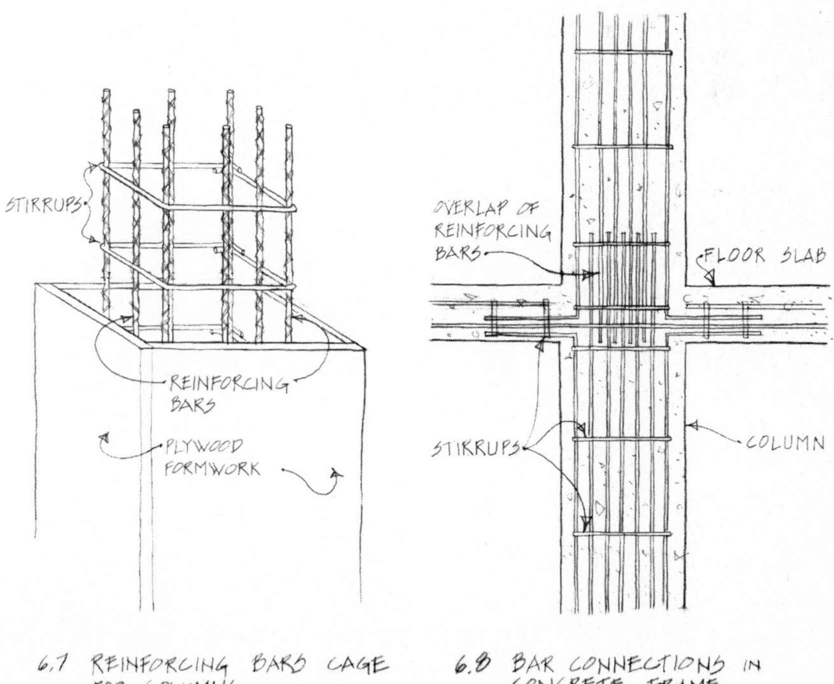

6.7 REINFORCING BARS CAGE FOR COLUMNS

6.8 BAR CONNECTIONS IN CONCRETE FRAME

tied to those coming out of the columns and the thin concrete floor slabs are then poured over the horizontal planks, with the chairs allowing the concrete to flow under the floor bars, which are thus completely encased. The procedure is continued floor by floor, dismantling the forms for the columns and the floor planks after a few days and reusing them to build the upper floors.

Once the concrete has completely set, the structural frame of the building consists of the columns, distributed more or less on a rectangular pattern, and of the floor slabs that, besides carrying the loads on the floors, act as horizontal beams to produce frame action between columns and floors in two directions. The length of bars lapped at the top and bottom of the columns (Fig. 6.8), the continuity between columns and floors due to the set concrete, and the tieing of the reinforcement of the columns and floors, create a structure that works always "together." If one part of a floor is loaded and bends down, the adjoining parts of the floor and the columns will also bend and help the loaded floor to carry its load (Fig. 6.9). Such a framed structure is, in a sense, most democratic, since each one of its elements helps the carrying action of every other element. This action, which is due to the continuity of the structure, diminishes as one moves away from the loaded area, just as in a family distant relatives are inclined to help less than parents, brothers, or children.

The reinforced concrete system just described gives rise to *flat-slab construction,* which is ideal for housing due to the limited gravity loads required by building codes, the small spans between columns, and the simplicity of running pipes and ducts along the flat underside of the floors,

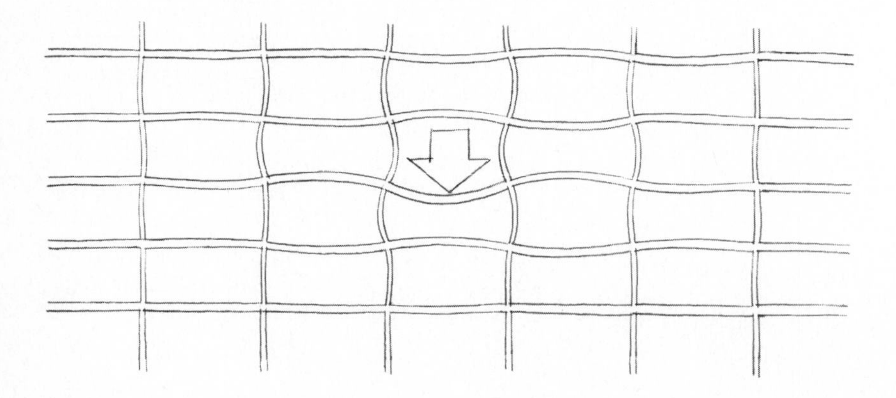

6.9 CONTINUOUS DEFLECTIONS OF FRAME

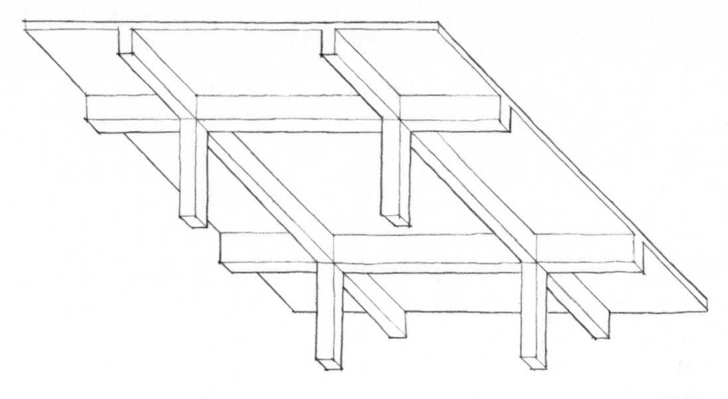

6.10 BEAM-SUPPORTED FLOOR SLAB

without having to duck horizontal beams. For buildings with longer spans or carrying heavier loads, the columns are connected at their tops in two directions by horizontal beams, which in turn support the floor slabs (Fig. 6.10). These are conveniently poured in thicknesses of up to twelve inches and reinforced with a grid of bars running in directions at right angles to each other, so as to act as beams in both the direction of the width and the depth of the building. When the required floor depth exceeds twelve or at most fourteen inches, floors are built as *waffle slabs.* These are created by placing on the flat formwork of planks for the floor square pans of steel or plastic, called *domes,* spaced at regular intervals in a rectangular pattern leaving gaps between them in two directions (Fig. 6.11). Reinforcing bars are set on chairs along these gaps and the

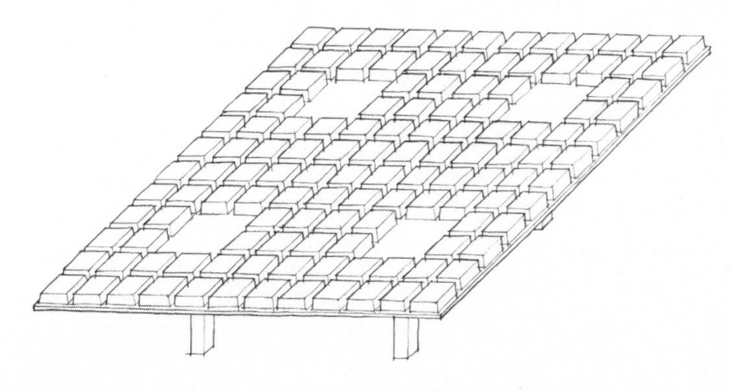

6.11 DOMES FOR WAFFLE-SLAB FLOOR

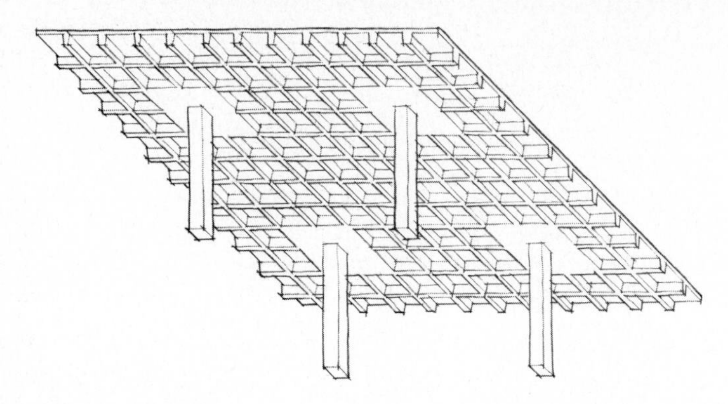

6.12 WAFFLE FLOOR

concrete is then poured to a depth a few inches greater than the height of the domes. Once the concrete has set, the formwork is lowered by knocking off its supporting struts and the domes fall off, since they have been previously greased and do not stick to the concrete. While the upper surface of the floor is flat, the underside presents beams in two directions, as wide as the gaps between the domes. This system of perpendicular ribs gives the underside of the floor the appearance of a waffle (Fig. 6.12).

The actual slab of a waffle floor, spanning between the close ribs, can be made quite thin, not more than a few inches. A large amount of concrete is thus saved, which would have been otherwise misused—located in the lower part of the floor, where tension develops (as it does in the lower part of a beam). The domes can be reused many times and the formwork of wood for the floor is as inexpensive as in a flat slab, since it is still built by means of flat planks. Waffle slabs can carry substantially heavier loads than flat slabs and are used mostly in industrial buildings, although their appearance is so interesting that some architects use them exposed in other types of buildings as well.

A concrete frame cannot be as high as a steel frame, but can be high enough to constitute the skeleton for a high-rise. The outer walls of such buildings are curtain walls, which do not carry loads. They can be made of light-metal window frames and contain a large amount of glass or they can have wall panels (with window openings) made of prefabricated concrete or blocks, similar to large bricks, glued with mortar (Fig. 6.13). Actually, the concrete blocks are strong enough in tension and compression to be used as bricks in load-bearing walls. The highest building erected so far with load-bearing walls of concrete blocks is the hotel at the Disney World in Florida, which is 192 feet high. However, high-strength plastic epoxy resin was used in its construction rather than mortar.

Prefabricated Concrete Buildings

One of the disadvantages of concrete construction is the time required for the concrete to set, which makes the erection of a frame relatively slow. Since money to erect buildings is usually borrowed from banks at high rates of interest, any procedure that speeds up construction reduces cost. This requisite has introduced during the last few decades methods of prefabrication in concrete construction.

There are two basic methods of concrete prefabricated construction—one by beams and columns, and one by slabs. Both very much resemble the way children put together "buildings" of sticks or packs of cards. The "sticks" method consists in fabricating in factories columns and beams of reinforced or prestressed concrete, which are "cured" at high temperature so that the material can set in a few hours rather than a few days. This pouring and curing can be done while the foundations of the building and

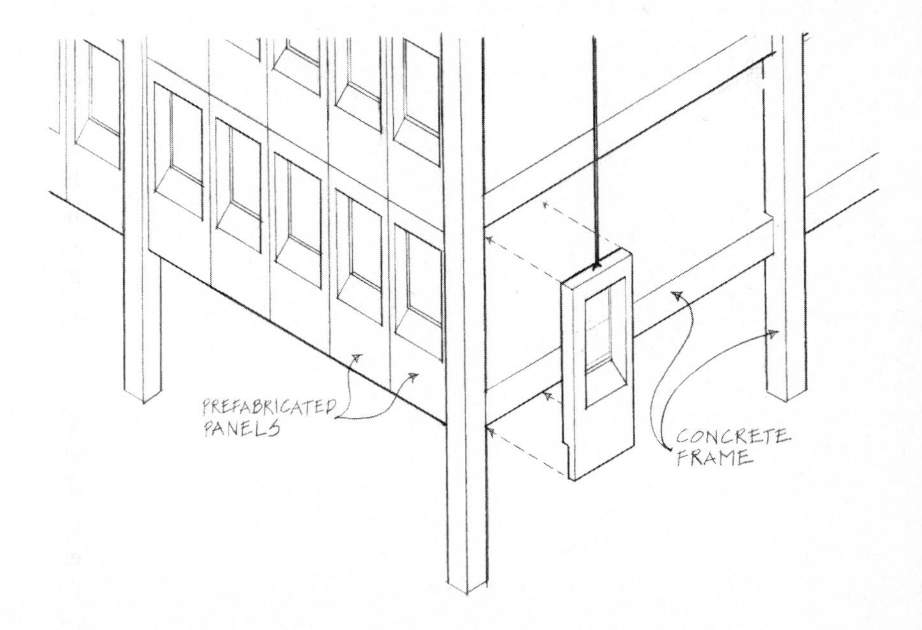

6.13 PREFABRICATED CURTAIN-WALL PANELS

other preparations are made ready so that time is saved both in the fabrication of the structural elements and the scheduled erection of the building. The prefabricated components, which can be shaped in a variety of ways at the pleasure of the architectural designer, are of very high strength due to the controlled curing process. Once ready, they are transported to the site and joined together to make a concrete frame either by pouring high-strength concrete at their joints, thus encasing the reinforcing bars left sticking out of them, or by prestressing tendons that pull together two or more adjoining elements (Fig. 6.14). A frame of such prefabricated elements can be erected almost as rapidly as one of steel and has become quite popular in countries in which the specialized manpower needed to execute the joints is not as expensive as in the United States.

By far the most magnificent applications of this method were Pier Luigi Nervi's airplane hangars built in Italy in the 1940s (Fig. 6.15). Measuring 1,080 feet by 427 feet, these hangars consisted of prefabricated reinforced-concrete struts (or trusses), connected into arch shapes by welding the bars sticking out of the struts and by concreting the joints between the struts. Two sets of such arches, at an angle to each other, created a

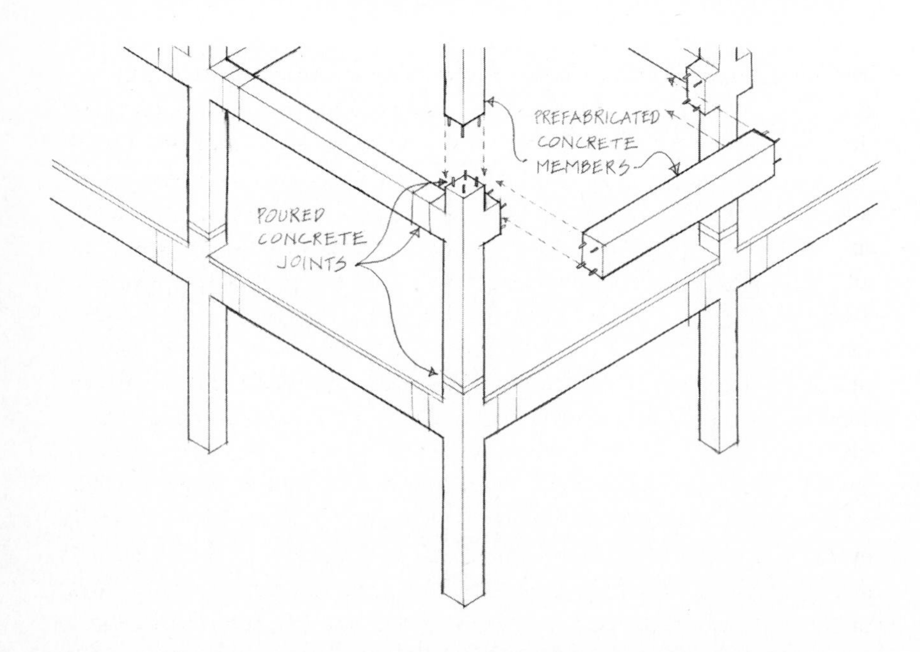

6.14 PREFABRICATED CONCRETE FRAME

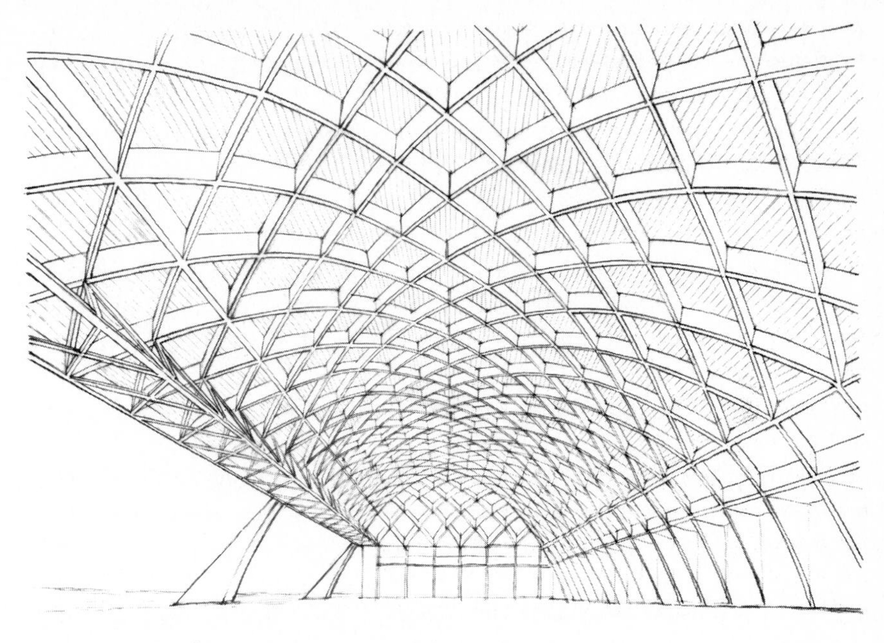

6.15 NERVI'S AIRPLANE HANGAR

curved surface, supported by four large arched buttresses at the corners of the rectangular plan and sometimes by two additional ones at the middle of its long sides. (Such a skeleton is commonly made out of wood or of steel in the United States and abroad and is called a *lamella* roof.) Since only small cranes, capable of lifting not more than two-and-a-half tons, were available to Nervi, he had to lift in place and connect many hundreds of struts at a height of forty feet above gound and then fill the openings between the struts by means of concrete slabs. The main criticism of this construction at the time was that it did not make use of the most favorable property of concrete, that of allowing monolithic construction. Nobody believed that the joints could make the structure as strong as a continuously poured roof. Unfortunately a most drastic experiment proved only a few years later how unjustified this criticism of Nervi's joints was. When the Germans retreated from Italy during the last phase of World War II they dynamited the six arched buttresses of these roofs, which crashed down to earth from a height of forty feet. No concrete structure had ever been submitted to such a dynamic test, but even so Nervi's roofs remained intact, except for a few among many hundreds of joints. The lesson, although painful to Nervi, was not lost on him and led him to even more daring prefabricated buildings later on, as we shall see in Chapter 11.

The second method of concrete prefabrication was perfected in France and other European countries after World War II and has been patented in a great number of slightly different forms. Essentially it rests on the concept of putting together a building like a pack of cards. The "cards" are reinforced concrete slabs, poured and cured on special tiltable flat boxes of steel. These boxes can be heated by steam running in pipes attached to their underside and to the hinged steel tops that close the boxes after the concrete has been poured (Fig. 6.16). After the few hours required for curing, the boxes are tilted to a vertical position and the slabs are grabbed by a crane that brings them to rest in the factory's yard, where many can be stood on end, one next to the other, without using too much yard space. Simple slabs of the type described are manufactured to be used as floors and interior walls, while more complex prefabricated slabs, incorporating a layer of insulation and openings for windows, are used for the outside walls of the building. The reinforcement of all slabs sticks out so that the slabs can be joined together by pouring concrete in the horizontal and vertical joints, thereby producing a monolithic structure which performs also the functional duties of outside walls and partitions (Fig. 6.17). The partitions are, of course, load-carrying walls, while the floor slabs are capable of carrying loads due to their two-way reinforcement similar to that in the slabs of flat-slab buildings.

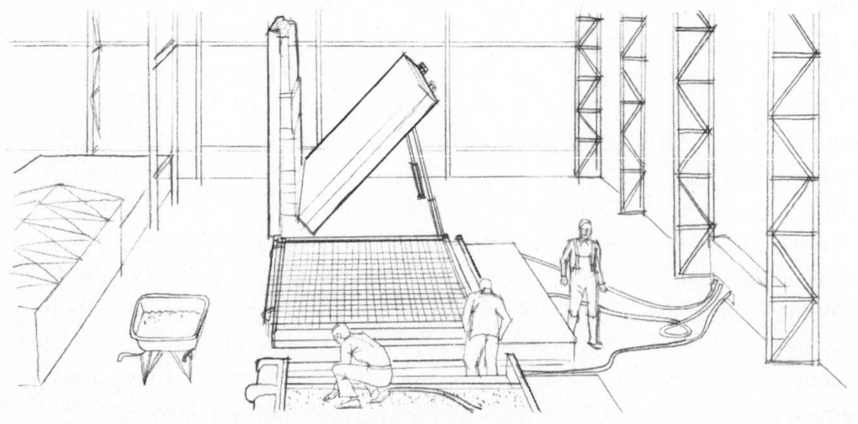

6.16 PREFABRICATED CONCRETE SLAB MOULDS

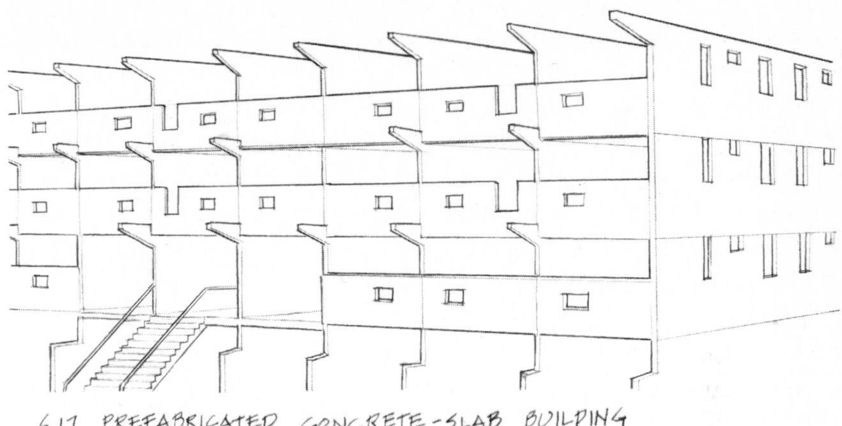

6.17 PREFABRICATED CONCRETE-SLAB BUILDING

One of the main reasons for the economy achieved by the slab-prefabrication method is that electrical conduits, water pipes, and air-conditioning ducts can be set into the forms before pouring the concrete so that the mechanical systems are incorporated in the structural system. Moreover, all work is done inside a factory and at grade level rather than in the open air and at increasing heights, saving time and providing greater safety and comfort for the workers. Buildings of up to thirty stories have been thus built in France, the USSR, England and the Scandinavian countries of Europe as well as in Japan and other technologically advanced countries. Notwithstanding the support of the federal government, the slab method of prefabrication that has been so successful in building inexpensive housing abroad has not become popular in the United States. This failure is due mainly to two quirks of the American mind. On one hand, both the architects and the presumed occupants of such housing have consistently rebelled against the "standardization" and "monotony" of their appearance, even though the dimensions, space distribution, and composition of such houses can be varied so widely that prefabricated-slab buildings are often less monotonous than our enormous low-income housing developments. The incredible construction by the prefabricated-slab method—in record time—of entire suburbs in Bucharest, housing 200,000 to 300,000 people, proves this, as do similar buildings in the Soviet Union. On the other hand, part of the economic advantages achieved by the incorporation of the mechanical systems in the slab has been lost in the United States because the construction trade unions refuse to allow it. They are fearful of losing lucrative positions for their workers to easily trained, unspecialized labor. Thus, some deeply rooted ideas, difficult to eradicate because not entirely logical, have limited the application of these modern methods of housing construction in the United States, where to this day the scarcity of housing is all-pervading. Lest it should be felt that Americans are alone in this prejudice, it might be instructive to note that in the city of Milan families on welfare refused to occupy apartments in concrete-slab buildings. They "could not stand living in and looking at buildings which had all identical windows." A comparison with the housing habits of the Patagonians and other primitive tribes shows that once the basic needs of humanity are served, other factors come into play. People will not be forced into buildings that do not attract them. One of the great challenges of architecture today is to join technological and economical requirements in innovative and winning designs.

7 Skyscrapers

The High-Rise Building

They were called skyscrapers, a name to exalt the human mind. They are now called high-rises—a matter-of-fact name that does not dare to reach the sky—and sometimes towers, a return to the Middle Ages and preoccupation with isolation and defense. In 1913 the first, the Woolworth Building, dared to reach fifty-five stories, soaring up 791 feet. It looked like a Gothic cathedral. Only eighteen years later the third, the Empire State Building, reached 102 stories and 1,250 feet and took on the appearance of a modern building. Most high-rises are square, but there are high-rises with from three to six sides, as well as round ones. Some of them seem to be made of ethereal glass and reflect the ever changing show of clouds and sun moving in a blue sky against a background of dark buildings. Others are as massive as fortresses of concrete. Some with the elegance of great ladies are clad in most expensive materials, like bronze. Others have the poise of distinguished matrons, dressed with slabs of black granite. Some are lily-white, with surfaces of anodized aluminum or treated concrete. Most exhibit in their facades the rigidity of geometrical forms, but some culminate in curlicued tops, reminders of Chippendale furniture. Some are textured and balconied, others as smooth as a wall of satin. Some proudly show on their surfaces bold structures of unrusting steel or of smooth, sandblasted concrete. Their roofs are flat, for the most part, but some are inclined at steep angles. A few have curved, sloping facades of dark solar glass, which act like the distorting mirrors of sideshows. Some consist of a single slender tower, others of two, sometimes identical like twins, sometimes of different shape or height, looking at each other in permanent challenge.

All high-rises contain offices, but some have stores at street level, hotels above them, offices higher up and apartments toward the top. Some "generously" give parts of their site to the public, to be enjoyed as gardens of trees and sculptures, in calculated exchange for the advantages of greater heights conceded by city planning commissions. Others enclose in their interior immense, glass-covered atriums reaching hundreds of feet, available to the public for the same hidden reasons and advantages. Some are built on "air rights" over slightly older buildings or behind them or up from their courtyards. Some contain such a variety of occupancies that one could be born and live a fairly varied life in them, leaving only at the end in a casket.

High-rises have changed the appearance of many a city, creating new skylines as impressive as mountain ranges. Even in socialist countries their location is carefully studied in relation to their effect on the landscape. They are a dream to the engineer and anathema to the city planner. Volumes have been written about their influence on the life of man, both from a physiological and a psychological point of view. They have grown suddenly all over the world. In the capital of one of the European socialist states, the visitor is shown before anything else *the* skyscraper—seventeen stories high. In a small town in Tuscany the citizens, oblivious of their medieval glory, point with pride to *their* skyscraper of six floors. Beyond any considerations of economy and energy savings offered in its defense, the skyscraper represents the fulfillment of one of man's aspirations, first voiced in the Bible.

The same compulsion that sent Mallory to his death only a few hundred feet from the top of Mount Everest drives men to erect taller and taller buildings. They matter not because of their absolute height, but because they overcome the height of other buildings. In the flat countries of the north, towers are built just to look down at the plains, and this in the age of the airplane. Their purest expression has not been achieved yet but has already been suggested: sculptures as high as skyscrapers, existing only to be looked at. The race is on to build the first high-rise with 150 floors. Frank Lloyd Wright, always a visionary and the wittiest of architects, designed, probably as a joke, a building one mile high.

But we should not be taken in by the skyscraper phenomenon. It is only seventy-five years old and could well be a passing whim of the mind of man. Posterity will tell whether, like so many other buildings, they will be demolished when they become economically inefficient or whether they will remain as permanent monuments to our dreams and folly. For the time being their favor is not diminishing. In 1980 twenty-seven of them were growing together in the center of Manhattan, healthy again after a few years of neglect due to temporary economic conditions. The growth of a high-rise is a civic drama of imposing magnitude. Let us understand it and, possibly, enjoy it from its very first act.

The Birth of a Skyscraper

If birth is counted from the day of conception, as in the Buddhist canon, the inhabitants of a large city are never aware of the birth of a skyscraper. Its conception is in the mind of a professional real-estate developer or of a corporation president years before the first step for its construction is undertaken. The search for a proper site starts only after a long series of thorough economic studies based on the requirements of the building and on the financing of its erection and operation. High-rise buildings may be built by developers to be sold to a client immediately upon completion or to be leased for long periods of time to a corporation. Banks must indicate the exact conditions of their mortgages; the city must express an interest in the building and specify the concessions, if any, it is willing to make beyond the rules and regulations of its building code.

At long last agreements in principle are reached on all these complex and interrelated matters, after a successful search for the site has been completed. With luck, a clear site may have become available, satisfying all the conditions. Most often the site had to be "put together" by buying in utmost secrecy a number of small adjoining sites, so as not to alert their owners lest they should increase their demands exorbitantly.* The owner of the last tiny corner-plot needed to complete a site has been known to hold up construction for months in order to obtain compensation ten times higher than the real value of his land. Sometimes this tactic fails. A few skyscrapers in New York City have been built around an old restaurant or pub, rather than accepting the prices demanded by these establishments.**

Once the site has been acquired, construction must start at the earliest possible date in order to save mortgage interests and to obtain early

* Putting together the site for the Avon Building in New York at 9 West 57th Street took years of stealthy assembling of lots.

** As is the case for the building labelled 900 Third Avenue in New York.

returns from the enterprise. It is at this moment that the first technical experts enter the game. Perhaps with the help of a computer program they decide, on the basis of many factors, the optimal size of the building, its height, the percentage of the site area to build on, the type of occupancy or occupancies, the size of the rooms, the number of elevators, their type and speed. The output of such programs has already played an important role in the search phase of the enterprise but becomes central to it now. At the same time an architect is chosen if he was not already a member of the team, and starts contributing his ideas to the realization of the building. He cannot do this alone. He surrounds himself with a large group of experts: the structural engineer, the mechanical engineer, the construction manager, the environmental engineer, the cost estimator, the city-code specialist. The team prepares sketches and estimates which must win the approval of the owner and of the various city departments of transportation, fire, building, and planning.

As soon as the team feels that approvals are forthcoming, the first public act of the drama is unfolded: the excavation of the site begins. The wooden fence is built around it, decorated by signs advertising all the component members of the construction team, holes are left in it for the sidewalk superintendents, and the noisy, messy, but highly organized process of excavating starts. Of course, if the site contains older buildings, these must be demolished by charges of dynamite, by the wrecker's ball, and by the dangerous and patient work of men perched on top of the ruinous walls, pick in hand, destroying the structure under their own feet. The demolition remains are carted away, but seldom discarded, since old steel is used in the manufacture of new steel and old bricks, made stronger by time, may be worth even more than new ones.

At this point steel piles called *soldiers* are pushed into the earth all around the boundary of the site at ten-foot intervals, and wooden planks, known as *breast boards,* are set behind them to insure that the sides of the excavation will not cave in (Fig. 7.1). Then holes are punched for inclined steel bars, which will support the sidewalks or the adjoining buildings. Many problems can be encountered at this stage, since networks of pipes and sewers may run under the site. A subway line may even cross it. In this case the subway tunnel is spanned by heavy steel girders, which may be seven or more feet deep and will eventually support the portion of the building over them.

When the soil to be excavated is not hard—clay, sand, or weathered rock—the big front-loaded shovels scoop it out and load it on trucks for removal. Solid rock must be dynamited. The drilled holes in which the

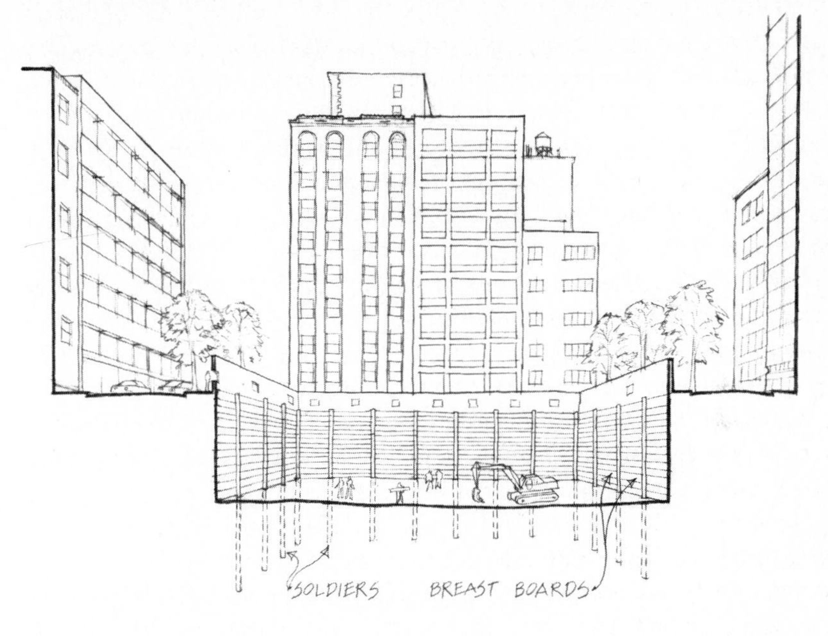

7.1 THE EXCAVATION OF THE SITE

dynamite sticks have been embedded are covered with *mattresses* of heavy chain mesh to check the flying debris. One whistle blast warns the workers that the explosives are in place, two blasts require that they take shelter, and three give the all-clear. The site is covered with seismographs —instruments used in measuring earthquake tremors—in order to assess the effect of the explosions on the adjoining buildings. These must often be underpinned with walls of concrete going deep into the ground lest the excavation and the explosions damage them. And, to play it safe, numerous photographs of the foundations of the adjoining buildings are taken to disprove any later allegations of damage.

Sometimes a deep excavation may reach soil capable of sustaining the enormous loads on the footings of the high-rise columns. At other times the soil continues weak to such a depth that it must be consolidated. This is done by hammering into the soil steel or reinforced concrete piles, reaching down to lower layers of stronger soil. *Pile drivers* do the job by dropping a heavy mass repeatedly on the top of the piles, to the desperation of the neighbors, who must endure this jarring noise for weeks, sometimes months. (For their peace, in New York City an ordinance forbids this work to start before eight in the morning or to last after six in the evening.)

There was a time when skyscrapers could be built only on solid rock. This explains the flowering of skyscrapers in Manhattan, which is mostly a rocky island. Modern soil techniques permit now the erection of high-rise buildings on the weakest soils. The John Hancock Insurance Company Building in Chicago, one hundred stories tall, is founded on soil so weak that its four corners had to be supported on numerous caissons—steel tubes three feet in diameter and filled with concrete—some of which tended at first to sink into the ground. The twin towers of the World Trade Center in New York City, the second tallest building in the world, rest on a soil of low consistency that was consolidated by a special chemical process devised by an Italian company. Although these processes are costly, they make it possible to erect a modern tower almost anywhere.

Once the work of the wreckers and the foundation contractors is finished, at long last the erection of the building starts. Certainly many months and at times years have passed since the idea of a skyscraper first glimmered in the developer's mind. Not that this time was wasted. Quite the opposite. All modern buildings are erected at present by the *fast-track method,* which consists in starting the construction of a building *before* its design is completed. There was a time when no construction would start until the last detail of the project had been designed and accepted by all concerned. Today these approvals would mean a delay of a year or more and such a delay is unacceptable due, among other things, to inflation, which pushes up costs of materials and labor. The prolonged mortgage payment periods and the delay in receiving the first income from the building are other considerations in favor of fast-tracking.

Fast-tracking is facilitated by the design procedure followed by the structural engineer. He first produces a *schematic design* based on architect's sketches in which the main components of the structural system are defined, such as the material (steel or reinforced concrete), the pattern and spacing of the columns, the type of floors, and the main wind-bracing system. Through experience this part of the design can be done in a few weeks, and approximate methods of estimating the quantities of required materials to within ten to fifteen percent, based on this scant information, have been devised. As soon as the schematic design is accepted by the architect and owner, materials to be delivered months later are ordered and contracts for their manufacture are signed. The structural engineer can then proceed with a more detailed *preliminary design* and cost estimate and, eventually, with the *design documents.* These consist of the *working drawings,* containing all the details of the structure, together with a thorough set of prescriptions for the manufacture and erection of the skeleton. These *specifications* constitute a document of the utmost technical and legal importance. The design documents are submitted to the building department for their official approval and are the basis for the *bids* to be taken and the prices to be paid. Open bids invite any contractor willing to participate, while limited bids are open only to contractors known for their capacity and integrity by the architect, the owner, and the other members of the team. The *contract documents* comprise the final drawings, whose preparation may cost seven or eight thousand dollars each, and the specifications. Behind them is all the computational work done on the hand calculator or on the electronic computer and based on the engineer's experience and the restrictions of the prevailing codes.

At this time the contractor winning the bid, who becomes there and then an important member of the construction team, must prepare more detailed drawings at a larger scale, the *shop drawings*, which are used by the fabricators in the shops to manufacture the separate elements and by the workers on site to put the building together.* These ultimate drawings are reviewed and finally stamped for approval by the structural engineer. It would seem that after all these checks and counter-checks few mistakes if any should ever occur. But people are fallible, and last-minute modifications do occur before the building finally goes up.

Shop drawings in hand, the superintendent of construction directs the various operations. The first is to pour at the exact location of the columns the footings on which the columns will rest. Next the steel columns manufactured in the fabricator's shop, with the holes for the high-strength bolts exactly located, are connected to the footings by bolting them to their protruding threaded bars until the forest of columns of the lowest floors is in place. These are then connected by the *main beams*, between which secondary or *filler beams* are often set to reduce the spans of the floor. Floors consist of reinforced concrete slabs poured on ribbed steel plates or *decks* (Fig. 7.2) which, supported by vertical struts during construction, serve both as forms for the concrete and as steel reinforcement instead of bars. Once the concrete sets, the struts are knocked down and the floor becomes a ribbed slab reinforced by the steel deck. The next set of columns is similarly erected together with the next-floor slabs and the skeleton grows, usually at the rate of a complete floor (columns and slab) every three or four days.

During this operation steel derricks—small, easily demountable cranes—are used to lift the columns and beams and set them in place, after

* A perfectly analogous procedure is followed in erecting a concrete high-rise.

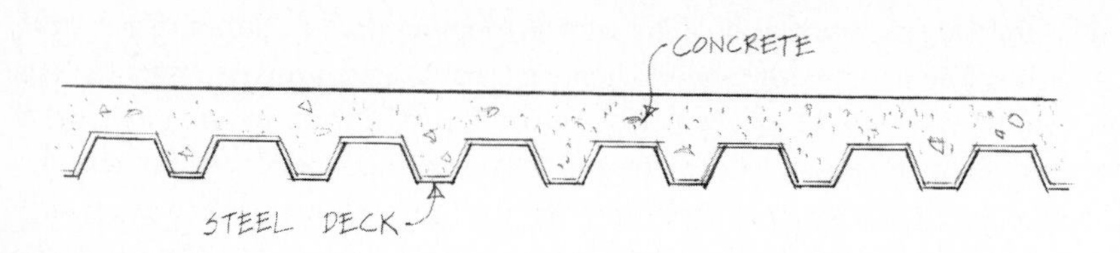

7.2 STEEL DECK FOR A CONCRETE FLOOR

which the derricks pull themselves up to the next floor by their own boot straps or with the help of the large cranes set along the periphery of the building to lift steel elements and concrete buckets to the high floors. For extremely tall structures only these derricks can do the work, since cranes reach at most 300 feet, and an elevator attached to the side of the skeleton lifts materials and people to the work level. It is particularly thrilling to see the iron workers, the *boomers,* walk on the narrow beams and tighten the bolts of the connections with the same nonchalance of the people looking up from the pavement many hundreds of feet below. Helmeted, carrying their tools on thick belts, these men *and* women perform an aerial ballet that would dizzy most of us.

When the steel skeleton has reached the projected height, the building is said to be "topped out." A flag is flown from the highest point and a celebration is held, where engineers and architects compete with the construction men in the appreciation of food and beer. But all the skill of the iron workers, the care of the superintendents, the rules of the safety agencies, and the supervision of the insurance companies do not eliminate entirely the possibility of accidents. Some members of the construction team may be absent from the celebration.

As the upper floors of the skyscraper are being built, the lower floors are completed by attaching the curtain wall to its exterior columns, so that work on the mechanical systems can be done in an enclosed space, made comfortable by temporary heat in winter. Vertical and horizontal pipes, electric conduits, and air-conditioning ducts are attached to the skeleton after it has been painted with coats of antirust paint. Eventually partitions and hung ceilings, which cover intricate networks of pipes and ducts, are set in place, together with the thermal insulation around the columns and the sprayed insulation on the lower surface of the steel decks which will make the building fire-retarded, as required by code, for the safety of its occupants. (The sprayed insulation, that formerly consisted of a compound of asbestos fibers, was responsible for the exceptionally high number of deaths by lung cancer among the workers specializing in this type of work. Asbestos is prohibited now and, at long last, new, safe materials have been substituted.) Partitioning and flooring are the very last jobs to be done, since they require the approval of the tenants and, at times, minor modifications to the structure.

There is seldom an opening celebration of a skyscraper. The tenants occupy their premises as soon as these become available, while construction continues elsewhere and work on details lingers for a long time, often years. But the skyscraper was born and grew strong and healthy and on schedule. The design and construction team feel deeply about this last accomplishment, but must thank for it an unsung hero of the entire performance, the construction manager. This is not a person but an entire company, charged with the task of making sure that the erection of the building proceeds smoothly, without any undue delay in the arrival of materials to the site and in their erection, in the various reviews by the engineers and the architect, in the approvals by the owner, the architect, and the city officials. This is a monumental task of bookkeeping, made easier and at times only possible by the judicious use of computer programs, capable of keeping track of the infinite number of operations

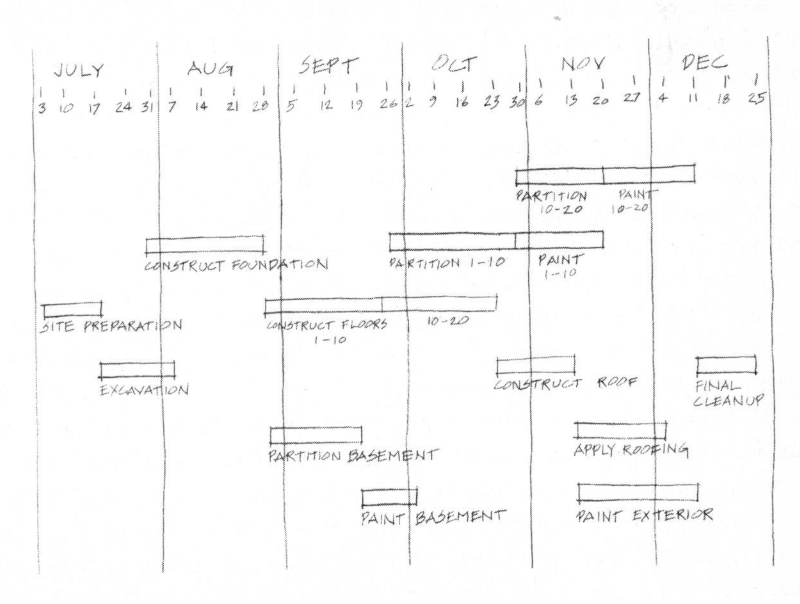

7.3 SAMPLE OF BAR CHART

and materials used in the erection of a large structure. These programs, which are updated daily and weekly and whose results are presented visually on *bar-charts* (Fig. 7.3), indicating the beginning and the end of each operation, are the only guarantee that the work is proceeding correctly and on schedule. A good construction manager is a blessing to the entire team, and his advice and admonitions are followed with the greatest respect.

When we look at a growing skyscraper from the construction viewpoint, we forget that its composer and conductor, who at all times orchestrates the performance of the work, is the architect. He is the creator of the building and the leader of the team. To him goes the glory and the total responsibility, to the owner, the occupants and, above all, to the public. Architecture is today one of the most exciting and creative professions in the world and so complex that few become recognized as great architects at the end of their career. Artist and technician, leader of men and expert in the everchanging fields of finance and politics, the architect is perhaps the greatest humanist in our complex and at times chaotic societies.

The Skyscraper Structure

The skyscraper is a triumph of American design and construction engineering. The key to its stability is a resistance to lateral wind or earthquake forces, which grow dramatically in magnitude with the building's height. For moderate heights of thirty or forty stories stability can be assured by the frame action of beams and columns rigidly connected at their joints. Reinforced concrete creates these rigid connections by its very monolithic quality. Steel is given rigidity by bolting the connections with high-strength bolts or by welding. Thus in 1883 Chicago engineer William LeBaron Jenney conceived and built the first building completely framed in steel, the Home Insurance Building, using all types of beam and column shapes. Only twenty years later the first high-rise made of reinforced concrete, the Ingalls Building in Cincinnati, was erected to a height of sixteen stories. The frames of these buildings were not only capable of channeling the gravity loads to the ground, but of resisting the wind forces with a relatively small sway, thanks to their masonry walls that added materially to their lateral stiffness. The wind was always the essential obstacle to the great construction heights made possible by the high-speed elevator invented by Otis in 1857. It is easy to understand why.

The weight of the structural materials needed to build a floor is the same whether the floor slab and its supporting beams are erected at the tenth or the one hundredth story, since each *floor* carries more or less the same loads. Columns, however, must carry the weights of all the floors above them. Therefore the lower floor columns carry much larger loads than those at the top. Actually the topmost columns carry only the load of the roof and their own weight; those of the next floor down carry twice as much load and, as the loads accumulate down the building, the bottom columns carry the weight of all the floors of the building. In other words, the load on the columns increases with the number of floors of the building, and their weight must vary in the same proportion. But both the wind forces *and* their lever arms increase with the height of the building. Their product, that measures the tendency to turn the building over (see Chapter 5), thus increases with the square of the height. The Sears Tower, about twice as tall as the Woolworth Building, must resist wind effects four times as large as those on the Woolworth Building. With the high strength developed by both modern steel and concrete, resistance to these forces would not be a real difficulty were it not for the swaying due to the elasticity of these materials which must be limited to insure the comfort of the skyscraper occupants. For reasons of both safety and comfort buildings must be stiff. In the first skyscrapers horizontal rigidity was obtained by the use of closely spaced columns and deep beams and by filling the voids between them with heavy masonry of brick walls.

The key to the solution of lighter frames and partitions was found when it was decided that the channeling of the gravity loads to earth and the resistance to wind forces should be attributed to two *separate* structural systems. This simple concept gave rise to buildings with relatively flexible exterior frames and an inner *core* of stiff wind-bracing frames, inside of which ran the elevators and many of the pipes and ducts of the mechanical systems. The frames of the core, in addition to beams and columns, had diagonal bars which *X-ed* their openings (Fig. 7.4) and gave them greatly increased stiffness by working in tension and compression rather than in bending. Remembering how the longitudinal deformations due to pulling and pushing are minute as compared to the lateral deformations due to bending, it is not difficult to realize how the triangulated frames of the core can be light but stiff enough to resist almost all the wind forces.

Diagonals, of course, cannot be systematically used in exterior frames without cutting across windows or in interior frames without destroying open door space. Three of the four frames surrounding the core can

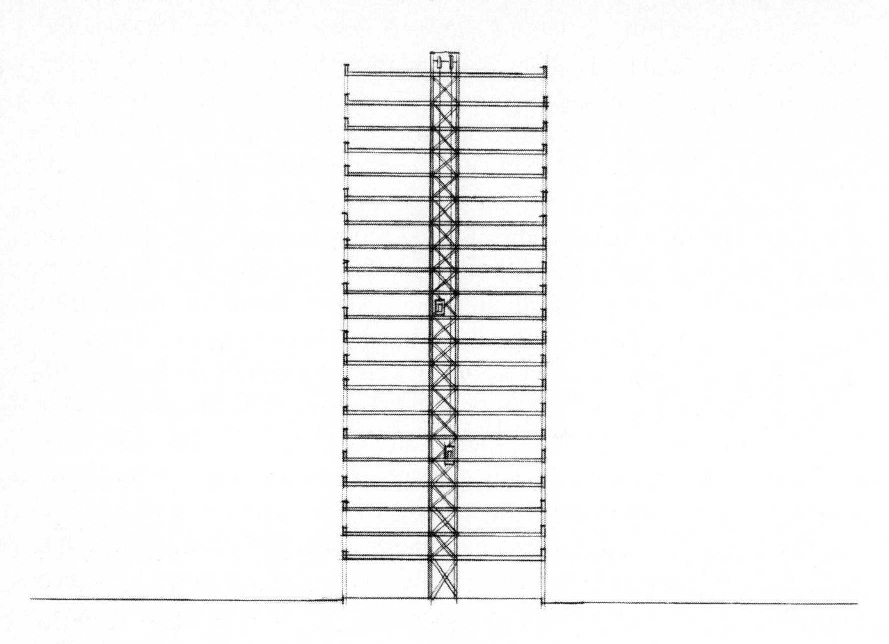

7.4 A TRUSSED CORE

be X-ed, and the fourth can be stiffened around the relatively small elevator door openings.

It was an easy step to realize that the cooperation of two different materials could achieve the same results with increased economy. Rather than stiffening the core frames by means of diagonals, one could fill their openings with thin, concrete walls. [Such walls have great stiffness in the horizontal direction, as one may find out by trying to slide in opposite directions the upper and lower sides of a piece of plywood or cardboard (Fig. 7.5).] It then becomes logical to build the entire sides of the core with reinforced concrete walls, thus obtaining an interior narrow, self-supporting, stiff tower, to which the light, outer steel frame is attached. To go the last step, one must be aware of the fact that in steel construction rigid or *moment* (bolted or welded) *connections* are costly (Fig. 7.6). They require specialized manpower and dangerous work at great heights. Their cost may represent ten percent of the entire cost of the structure. But, if the inner core were stiff enough, one could foresake the rigid connections between the beams and columns of the exterior frames and use much cheaper connections, which allow beams and columns to rotate one with respect to the other, as if they were hinged (Fig. 7.7). Such

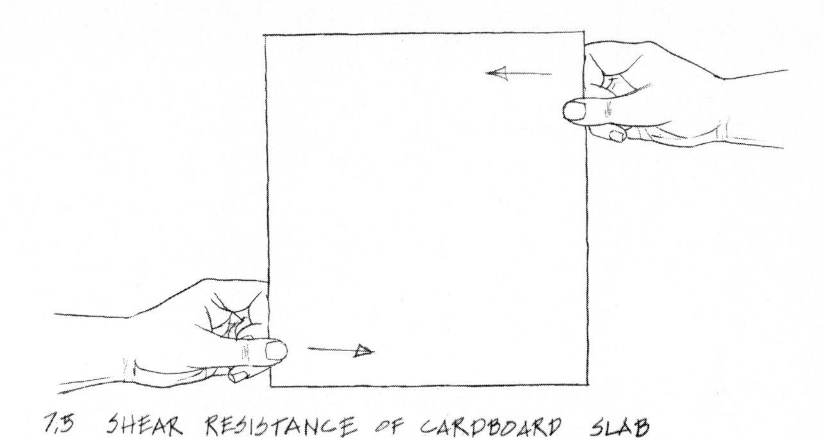

7.5 SHEAR RESISTANCE OF CARDBOARD SLAB

hinged, or *shear,* connections could not be used without a core, since the frames would collapse like a house of cards (Fig. 7.8), but they are economical and practical if the core stands up rigidly and the outer hinged frame leans on it. The separation of the two structural functions is now complete: the hinged frame carries the vertical loads to the ground and the core resists the wind forces.

The principle of using the walls of a concrete core, the *shear-walls* or the X-ed inner-core steel frames, together with a hinged outer frame has reduced the weight of skyscrapers very considerably. To realize the efficiency achieved by the core system, it is enough to notice that the Empire State Building, 102 stories high, has a total weight of steel of

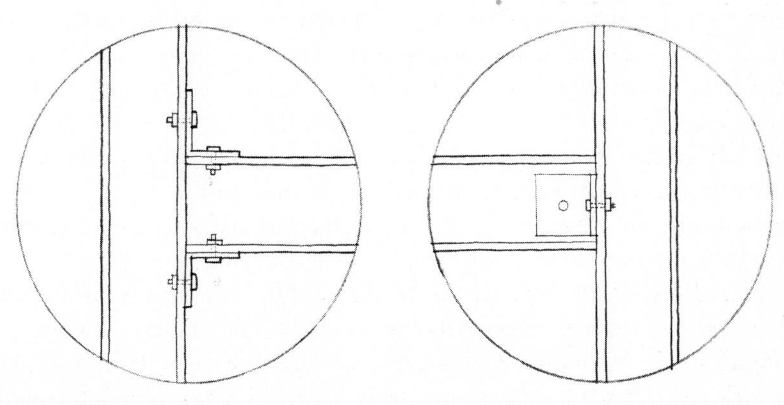

7.6 RIGID OR MOMENT CONNECTION 7.7 SHEAR OR HINGED CONNECTION

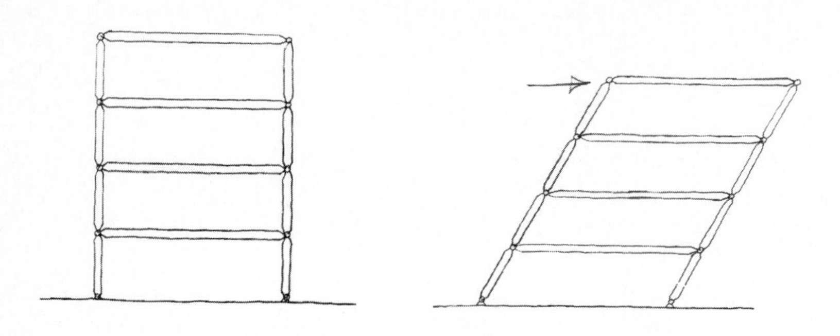

7.8 COLLAPSE OF SHEAR CONNECTED FRAME

42.2 pounds per square foot of floor, while the John Hancock Insurance Company Building in Chicago (Fig. 7.9), 100 stories high, has a weight of only 29.7 pounds. This is a saving of forty-one percent. It was achieved by a modification of the core concept in which the X-ed frames, instead of constituting the inner core of the building, are its outside frames, stiffened by five sets of enormous diagonals on the building's facade (Fig. 7.9). The John Hancock Building has thus done away with an actual core and utilized the outside walls to resist the wind. Of course, the diagonals block entirely the view of two windows at each floor, but a clever renting agent has made hay out of this obvious disadvantage. He transformed the diagonals into a status symbol by charging higher rent for offices and apartments which display diagonals out of their windows! A similar structural concept was used in the Avon Building in New York City (Fig. 7.10), where the X-ed walls are the narrow lateral walls of the building and resist the wind acting on the wider, longitudinally curved walls. In this building the wind on the narrow lateral walls is resisted by the frame action of the columns and beams without any need for an inner core. This "mixed" system using both X-ed walls and frame action could be used because the building has only fifty stories.

The new concept of an outer core reduces the skyscraper to an immense, hollow, cantilever beam, stuck into the ground and anchored to heavy foundations. The towers of the World Trade Center have outside columns spaced only three feet from one another and connected by deep outer beams, or *spandrels*. They act more like shear walls with small window-openings than as frames and they give the cantilevered towers

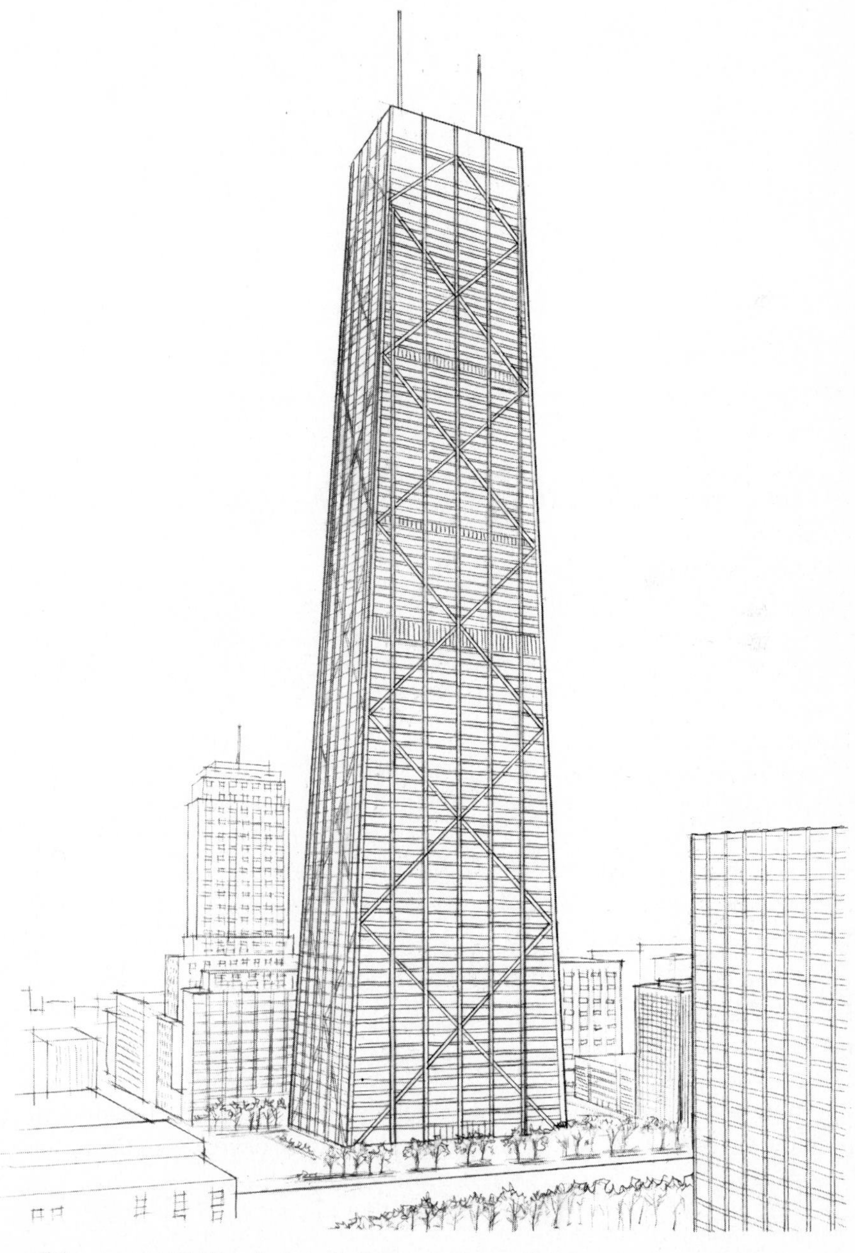

7.9 JOHN HANCOCK INSURANCE COMPANY BUILDING IN CHICAGO

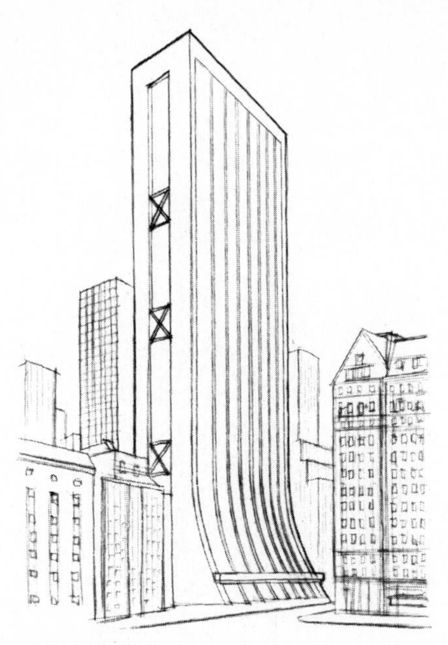

7.10 AVON BUILDING IN NEW YORK CITY

their needed lateral stiffness. Buildings like the towers of the World Trade Center or those of the John Hancock Insurance Company Building in Chicago are often called *tube buildings* since their outer walls act as the walls of a hollow tube. They are the most efficient structures designed so far against the wind and can even be combined into a "bundle of tubes," as was done by the Pakistani engineer Fazlur Kahn, one of the great engineers of our time, in the tallest building in the world, the Sears Tower. This consists of nine square tubes, each seventy-five feet square, erected one next to the other in a pattern of three squares by three squares and reaching different heights (Fig. 7.11). Although the Sears Tower is 1,450 feet high, the weight of its steel frame per square foot of floor is only thirty-three pounds. At the present time (1980) this is the most notable refinement of the tube concept and the greatest achievement in the field of skyscrapers.*

Reinforced concrete is lagging behind steel in terms of achieved building heights but has contributed interesting developments in the

* The reduction of wind-induced oscillations by means of tuned dynamic dampers is discussed in Chapter 3.

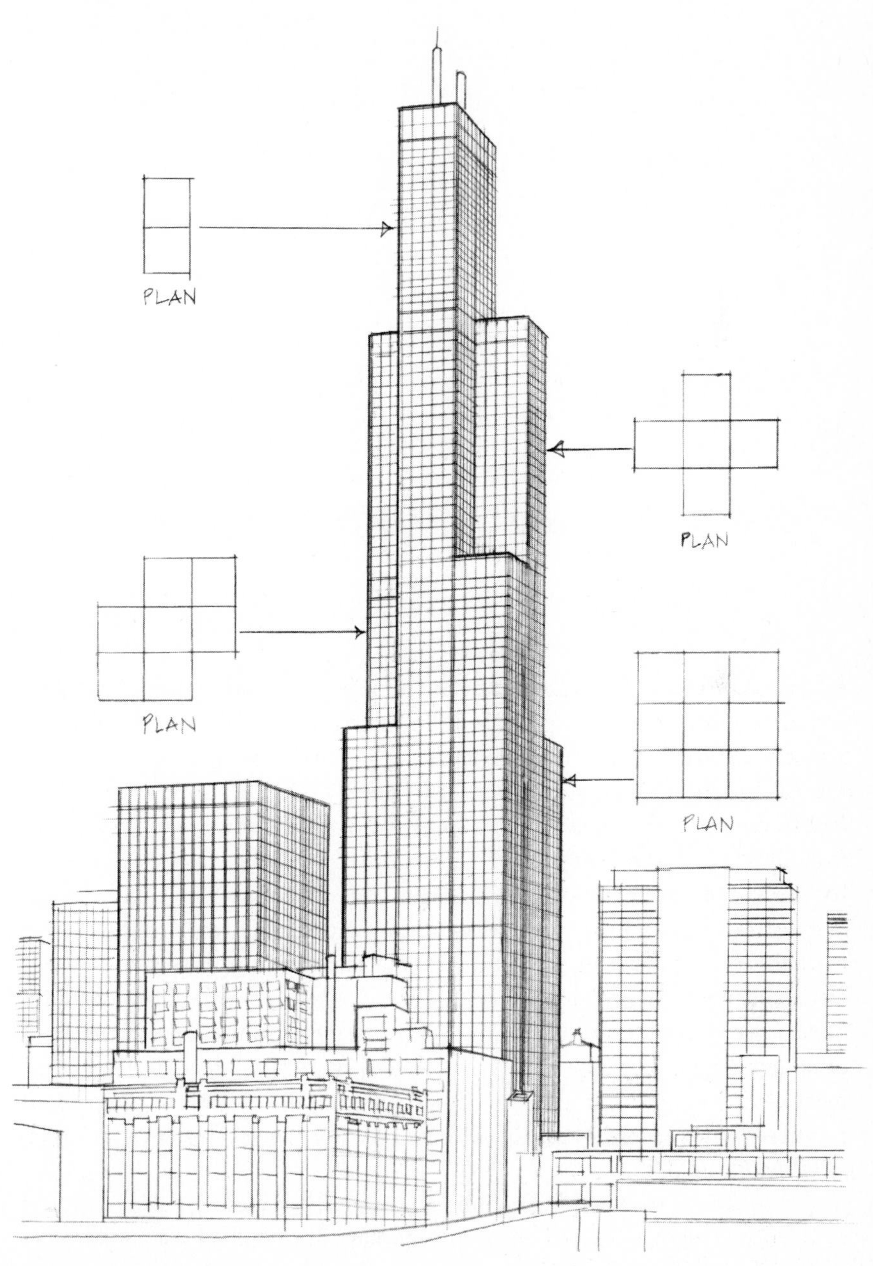

7.11 SEARS TOWER IN CHICAGO

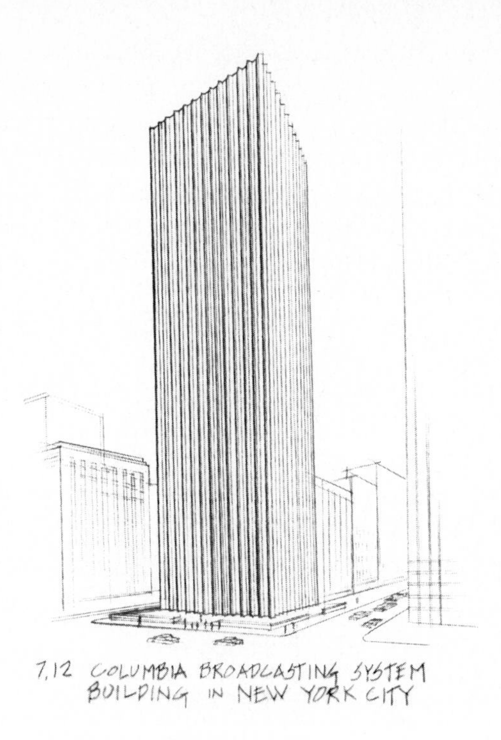

7.12 COLUMBIA BROADCASTING SYSTEM BUILDING IN NEW YORK CITY

field of skyscraper construction. In the reinforced concrete, forty-two story Columbia Broadcasting System building in New York City, which was designed by Eero Saarinen and engineered by Weidlinger Associates in 1962 (Fig. 7.12), wind forces are resisted by both an inner core and the shear wall facades of columns five feet wide and spaced five feet apart, connected by the rigid waffle floor slabs. Taller buildings in concrete have been first built in Brazil and then in the United States. The initial success of the Brazilian engineers was due to the fact that elevator requirements in their country are less stringent, so that the bottom concrete columns could occupy a larger amount of floor space, which in the United States had to be allotted to the elevators. With the increase in concrete strength obtained during the last few years, buildings of 750 feet and 800 feet have been erected in Brazil and in the United States, which at the present time holds the record with the 859 feet Water Tower Building, erected in Chicago in 1977.

Are skyscrapers a blessing for the densely populated areas of our modern metropolises or are they a dehumanizing fruit of technology for technology's sake? Are they the symbol of our overpowering interest in financial returns or of our spiritual aspirations towards the conquest of natural difficulties? Are they an expression of a madly materialistic culture or the realization of man's dreams of long standing? Are these enormous, air-conditioned beehives the ideal environment for modern man or are they a negation of our individuality and of our communication with nature? Whether we believe in skyscrapers or not, let us remember that the aspirations and realizations of man have taken different forms in different ages and that most certainly the skyscrapers will also disappear when the time comes for their demise. The impermanence of our structures is the best hope for our future, whether it will take us to live in space or underground.

8 The Eiffel Tower

The Symbol

The three largest steel structures in the world were products of an outburst of structural creativity in the last half of the nineteenth century. The Crystal Palace, erected for the 1851 Great Exhibition in London, covers the largest area; the Brooklyn Bridge, completed in 1883, has the longest span, and the Eiffel Tower, started in 1887, reaches the greatest height. All were made out of steel. The first two had utilitarian purposes. Sir Joseph Paxton's Crystal Palace not only served royally the Great Exhibition but, being demountable, was transferred afterwards to Sydenham, where it was used for sculpture, painting and architectural exhibits, and for concerts until 1936. (It was demolished in 1941, being an excellent guide for Nazi aircraft.) The Brooklyn Bridge, built to connect the two great boroughs of New York City, reflected the desire to increase real estate values in Brooklyn. The Eiffel Tower, on the other hand, is a case in itself. Built for the Paris Exhibition of 1889, it was meant to provide ostensibly nothing more or less than a magnificent view of Paris. Monsieur Eiffel extolled its utility as a military look-out, the salubrity of the air at its top, its uses as a laboratory to experiment with wind and gravity, but the innerspring to its construction was to demonstrate that France, 100 years after the revolution, was a leader of the technical world, capable of realizing the dream of a tower 300 meters high, almost 1,000 feet. It would be twice as high as the Washington Monument, at that time the tallest structure (Fig. 8.1). Whatever its motivation, the tower in itself was and is totally "useless" from a practical point of view.

8.1 THE EIFFEL TOWER

If we leave aside defense towers built by the Saracens all along the Mediterranean coast and similar defense towers of the medieval walled towns of Italy and France, almost all the other great towers of the past were built for spiritual purposes. They rose next to Renaissance churches and from the tops of Gothic cathedrals calling with their bells the prayer hours. As minarets they served and still serve to remind Moslem worshippers of their well-timed duties. They were also built as symbols of civic pride on or next to city halls. If they could be used to warn of an approaching enemy or bring citizens to important gatherings, so much the better. And they were tall: 200, 300 feet and more, all of them of stone, masonry, or bricks.

La Tour Eiffel acknowledges nothing so lofty or utilitarian. It was from the beginning a gigantic lark, an iron toy from which Parisians might admire their city, at the cost of one and one-half million 1889 dollars. And did it get hell from the defenders of the French tradition of beauty! Bouguereau, the academic painter; Garnier, the architect of the Opéra; Dumas, Jr., the novelist; Gounod, the composer; Leconte de Lisle, the poet; Prudhomme, the essayist; Guy de Maupassant, the novelist, all signed a resounding letter of indignant protest to the Minister of Public Works in an attempt to stop "the horror." Guy de Maupassant, never relenting, had lunch at the Tower as often as possible so as *not* to have to look at it. And he had a point, for the Tower so dominated Paris it could be seen from anywhere in the city, except from inside one of its restaurants.

If La Tour was an insult to the representatives of the "effete class," it was love at first sight for the people. Two million of them flocked to visit it during its first year. More than half of them reached its top. Thousands climbed the 1,671 steps before the elevators were open to the public. The crowds increased even long after the exhibition closed and, slowly but surely, their visits acquired new meanings. They went to look *at* the Tower as much as to look *from* it, to look inside, at its filigree of steel, as much as to point out the other monuments of their city. It became *the* symbol of Paris, the Mecca of all travellers, visited by far more people than Nôtre Dame or Sacré Coeur. And then it became, somehow, the symbol of France. Tourist posters ignored virtually all else. The "new" poets, writers, painters and musicians exalted its shape, its lightness, its infinitely widening significance. It became the subject of fifty-one paintings by Delaunay, the initiator of the dismembered visual image, who drew it simultaneously from top and bottom, right and left. Hitler vowed to destroy it. The Resistance fighters of World War II hoisted the tricolor from its top while the Nazi Panzer divisions were still battling American tanks in the streets of Paris. The Algerian rebels planned to dynamite it, as did their French right-wing military opponents, both in vain. By then the Tower had become like a mountain that always was and will always be. Paris would be inconceivable without it. Surpassed in height by many a modern utilitarian tower, by chimneys, antennas and skyscrapers, it is still the Tower of Towers. Of no other man-made structure can one say that it transcended technology to become a great human symbol.

The Builder of the Tower

No bourgeois is as bourgeois as a provincial Frenchman and Alexandre Gustave Boenickhausen-Eiffel (1832–1922) was no exception. He was the soul of respectability. A dutiful and bright student, he nevertheless flunked an entrance examination to the prestigious École Polytechnique and thus attended École Centrale, studying chemical engineering. His modest goal was to gain employment in the vinegar factory of one of his uncles. Luckily the factory failed. Gustave Eiffel became a structural engineer and, not incidentally, a rich man. By 1867 he had started his own iron fabricating concern and began to be both the designer and the contractor of his own works. As a designer he championed a new breed of engineers, the mathematical engineer, who relied totally on his calculations after initial tests on hypotheses, materials, and procedures. He was not flamboyant in either speech or behavior, but had the calm self-assurance of the scientist-engineer, who studies a problem first, chooses the most practical among its many solutions, and proceeds, without hesitation, to the realization of his idea. He cannot be said to have invented any revolutionary device, but he perfected so many that his works add up to a large number of original insights. All of them are economical in execution and ingenious in conception. He was so methodical and unflashy as to contradict the classical image of the genius. But a genius he was. What else can we call a man, who produced 14,352 square feet of drawings for the Eiffel Tower—drawings which called for 15,000 structural members and 2,500,000 rivet holes—and who put this immense jigsaw puzzle together without, as far as we know, a single error? A man whose every rivet slid smoothly into its respective hole? A man who erected the 8,000-ton tower without a single accident to its 250 workers, its *only* 250 workers? Who promised, and did deliver, the finished Tower in two years, two months and five days, on time for the opening of the Exhibition, although progress was delayed by inumerable squabbles with the supervising committee and several strikes by his workers? Who estimated the cost of the Tower at $1,500,000 and built it, making a handsome profit, at five percent below budget? Who was so sure of his success that he agreed to finance the construction of the Tower and to pay, outside of his contractual obligations, not only for all the workers' insurance but also for any damage claims by outsiders? And finally, a man who emerged triumphant from a financial scandal, and at age fifty-seven started a new and brilliant thirty-year career in the burgeoning science of aerodynamics? It has been said that genius is ninety percent perspiration and ten percent inspiration. The most amazing aspect of Eiffel's genius is that he does not seem to have perspired.

Gustave Eiffel did not come to the Tower unprepared. For twenty years he had built train stations, dome observatories, large department stores, churches, and above all bridges. All of his structures were made of iron. He designed and built all over the world, including Indochina, Russia, and Peru. Only one of his masterpieces is in the United States–the skeleton for the Statue of Liberty (Fig. 8.2). In each structure he introduced some new concept. The dome of the observatory in Nice was so supported that it could be rotated by hand. His bridge-pier caissons were

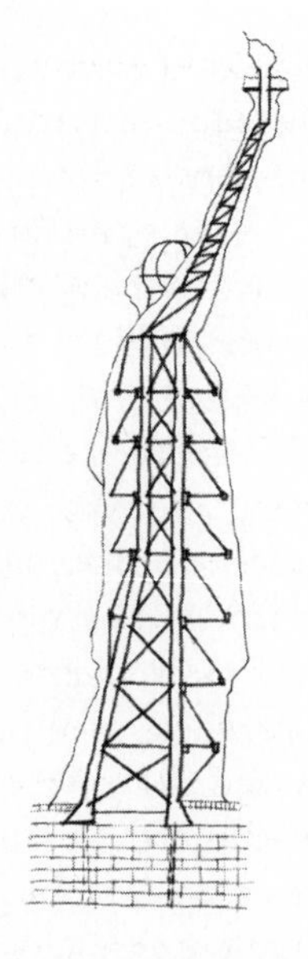

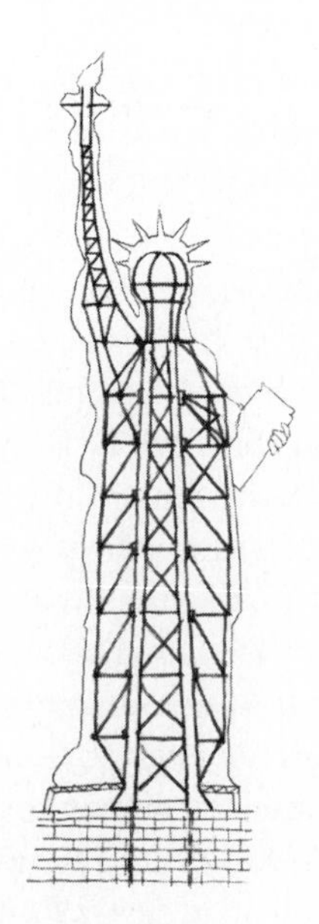

8.2 THE SKELETON OF THE STATUE OF LIBERTY

8.3 EIFFEL'S GARABIT BRIDGE

so perfect that no worker suffered from the bends. When bridge piers could not be built in a river, he launched his bridges from both banks with the help of temporary cables, as Eads had done before him (see Chapter 9), but over much longer spans. He was the first bridge engineer to realize and assess the importance of wind forces on tall bridges and to lace them with trusses (see Chapter 9) to minimize their effects. In addition he often laid the two iron arches supporting a bridge on inclined planes, one leaning against the other and supporting each other at their crowns to increase their resistance to lateral winds (Fig. 8.3). He spread apart the arch legs, as we do when we wish to resist a shove (Fig. 8.4).

The use of bridge caissons made him thoroughly familiar with soil conditions, and his mastery of steel came from an exact knowledge, personally acquired, of its properties. When his company bid on the Tower, he was ready technically, financially and politically. Knowing that no great civic work was ever built without the help of the politicians, Eiffel made their job simple. He indoctrinated them so well that by the time the bid papers for the Tower were ready only one company in France could successfully bid on it: Eiffel and Company.

8.4 Eiffel as The Tower (from Punch, 1889)

The Building of the Tower

A tower is a challenge to the wind. Its own weight and the weight of the people and the objects on it have to fight the overturning force of wind pressure, an enemy whose greatest blows have not been felt yet. This battle can be fought in two essentially different ways. The weight of the tower can be made so large that by brute force it will always counterbalance the wind. Or, more elegantly, one may make the surface of the tower so minimal that the wind has little on which to apply its pressure. Towers of brick and masonry had to rely on the first principle, but iron made the second possible. Eiffel, the master of iron and the first student of aerodynamics, relied entirely on the second method. He made his Tower a filigree of steel on which the wind has almost no grip: all pieces of the Tower, except some vertical columns, are an open lattice of light trusses through which the wind can blow. Eiffel also played safe. He assumed a wind, never yet encountered in Paris, of up to 148 miles per hour at the top of the Tower, diminishing to 105 miles per hour at its foot, or with an even speed of 134 miles per hour from top to bottom. It was easy to prove that, even ignoring the laced nature of its surfaces, the weight of the Tower could easily resist such wind forces. So much so that Eiffel could ignore in his design the weight of its 10,000 daily visitors, whose combined weight was at most a meager ten percent of the weight of the Tower.

The scheme of the Tower's design is simplicity itself (see Fig. 8.1). Starting at the top, four steeply inclined, square, boxed columns come down to the level of the second platform, 380 feet above the ground. The four columns are connected by windbracing X's of laced trusses, which transform them into a rigid obelisk of steel, inside which run, one above the other, the last two elevators from the second to the third platform at the top of the Tower. The slight curvature of the corner columns emphasizes the aspiration of the Tower towards the sky. This feeling is reinforced by two secondary columns placed at the middle of the Tower sides, also curved and meeting at an intermediate platform, where the two inside elevators stop to allow passengers to transfer from the lower to the upper.

The second platform introduces a strong horizontal connection between the four upper columns, tying them together at their base, and is supported, in turn, not by four but eight heavier columns. Four on the outer corners of the Tower continue with increasing curvature the slight curvature of the upper columns, while four on the middle of the sides are parallel to the corner columns. All eight end at the first platform 180 feet above the ground. Below, each corner of the massive first platform is supported by four inclined buttresses consisting of four straight columns, each inclined inward at an angle of fifty-four degrees. These enormous buttresses spread the base of the Tower to a square 420 feet on each side and give the visual impression that the Tower is solidly grounded against any wind blowing on its thinner upper part, the top platform being a square only thirty feet square.

The structure of the Tower does not end at ground level. Its 8,000 tons must be supported by solid foundations, which consist of four immense, inclined concrete pylons going into the ground to depths depending on the strength of the soil under each of the four inclined buttresses they support. While excavating the ground for these pylons and pouring their concrete, Eiffel had to make use of his experience with underwater caissons, since the water of the Seine seeped into the soil of the Champs de Mars, where the Tower is situated.

The final silhouette of the Tower is a visual representation of its structure: four massive buttresses inclined towards each other and prevented from falling by the powerful first platform, continued in the slimmer curved columns between the first and second platforms, and ending in the thin, almost verical columns of the upper 620 feet of the tower, which support the tiny, domed, top platform and the flag staff (now a TV antenna). The iron for this tallest of towers, not surpassed in height until 1929 by the erection of the Chrysler Building in New York, is so light that if the Tower were squashed into a plate the size of its base—four acres—it would be only two-and-one-half inches thick.

But if the scheme of the Tower is simplicity itself, the erection of its 15,000 parts presented problems never encountered in engineering before. To begin with, the four inclined buttresses at the foot of the Tower had to be temporarily supported during construction lest they fall inward before the erection of the first platform. To this purpose, Eiffel designed four gigantic wooden trusses which propped up the buttresses like the hands of giants (Fig. 8.5). Possibly the most difficult technical problem he encountered was that of the accuracy with which the four enormous buttresses had to meet the first platform at a height of 180 feet, so that the connections between the buttresses and the platform could be smoothly executed. If the inclination of the buttresses had been incorrect by only one tenth of a degree, the rivet holes in the platform and the buttresses

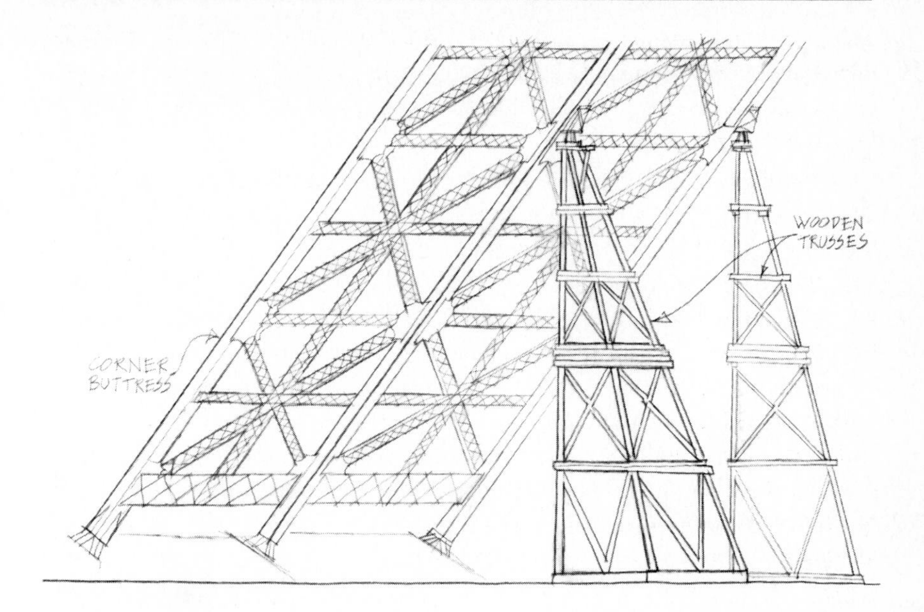

8.5 TEMPORARY WOODEN TRUSSES PROPPING THE CORNER BUTTRESSES

would have been five inches off and the connections could not have been realized. The constructive genius of Eiffel found a simple solution to this problem. He supported each of the four columns of the corner buttresses on a hydraulic jack, which could be operated by two men through a pump activating a piston. By controlling the amount of water in the cylinders of the jacks, the location of the bottom of the columns could be adjusted with great accuracy and then fixed by inserting an iron wedge between the column and its supporting concrete pylon.

An ultimate refinement in the location of the top of the four buttresses was obtained by an even more ingenious method. At the points where the four corner buttresses were temporarily propped up by the gigantic wooden trusses, Eiffel placed steel cylinders, each filled with fine sand and with a hole at the bottom, and four jacks capable of lifting the buttresses from the wooden trusses. If a buttress was too high, even by a fraction of an inch, it was lowered by letting the sand run out of the cylinder. If the buttress was too low, the jacks could lift it, matching the rivet holes in the buttresses to those in the platforms. A round bar was pushed through the holes to align them perfectly and then a rivet, heated on a portable forge, was inserted into the holes and hammered by hand from the side opposite its head, creating a connection between the two elements that became tighter when the rivets cooled and shrank. Of course, the structural drawings for each one of the elements had to be extremely accurate and their fabrication just as exact for the 15,000 pieces to fit as perfectly as they did.

The correct structural design of the Tower has been proved through ninety years of use. The only structural change ever required was the opening up and lacing of two sides of the box columns between the first and the second platforms to allow easier inspection and painting.

The incredible structural honesty of the Tower design had to bow to the aesthetics of the time in only two areas. Ten gingerbread arches of steel were built around the first platform for purely decorative reasons. Obviously contrary to the linear design of the structure, they were eventually taken down, with a substantial improvement in the Tower's appearance. Unfortunately, the second concession to the fashion of the times is still part of the Tower structure. Eiffel himself, afraid of the reaction of the public to the inclined buttresses supported by the horizontal square ring of the first platform, added to the four sides of the Tower, under the level of the first platform, four decorative arches that have no structural purpose whatsoever (see Fig. 8.1). He thought they might reconcile the Parisians to his daring design by reminding them of the familiar arches of their bridges. He thus gave the impression to the casual observer that the lower level of the tower was supported by commonly used arches, while these have no role but to fool the eye and detract from the strong action of the inclined buttresses leaning against the quadrilateral of the first platform. For aesthetic and structural reasons, these four fake elements should be dismantled, although by now they are thought to be an essential component of the Tower shape. The outrageous fakery of the "supporting" arches becomes obvious where the lower parts of these arches, after just touching the supporting straight buttresses, curve inward like the arches of a mosque. Destroying the linearity and simplicity of the Tower base, these "arches" are a detriment to the Tower's aesthetics and its magnificent structural honesty.

Cranes and Elevators

Once the Tower had been conceived and designed in all its details, two more problems had to be faced: how to lift the iron beams and columns to their final destinations and how to bring the estimated 10,000 daily visitors to the three platforms. These problems in mechanical engineering were somewhat outside the competence of Eiffel, although he had encountered the first in the construction of his large bridges.

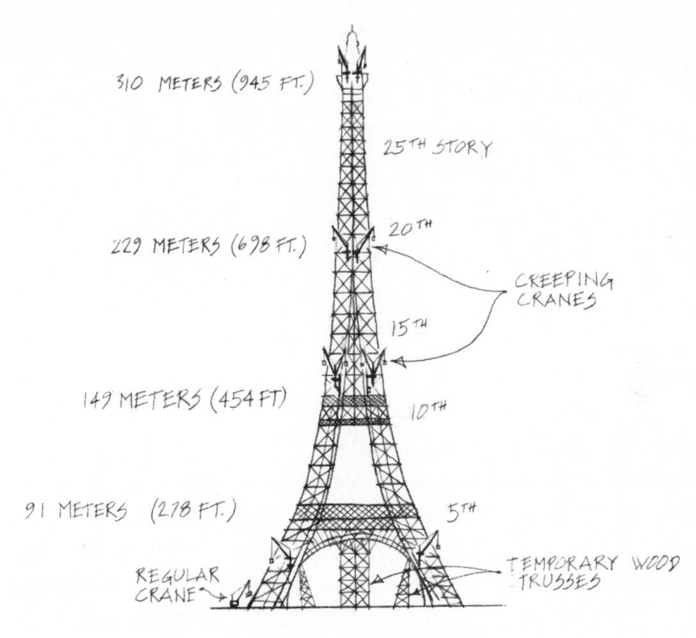

8.6 SCHEME OF THE CREEPING CRANES

Aware of the speed with which relatively light iron elements could be maneuvered, he limited the weight of the heaviest piece to three tons. This allowed the use of small cranes and winches and paid in terms of ease and speed of construction. The first pieces were lifted by regular steam powered cranes which had to be abandoned as soon as their reach was exceeded. At this point Eiffel introduced a system of cranes (Fig. 8.6), capable of *creeping up* the rails of the future elevators, set inside the bottom buttresses. The crane arms could rotate a full 360 degrees and cover the needs of an entire buttress. Once the first platform had been built, he used winches to lift the structural elements up to it and other winches to lift the additional elements from the first to the second platform. To move the steel components up to the top of the Tower he again used a crane capable of creeping up the central columns that were to serve as guides for the top elevators (Fig. 8.7). This crane was supported on a frame that allowed it to move up thirty feet and then proceed to another frame set above the lower frame. Thus the central crane, which could reach all four columns of the upper part of the tower, needed only

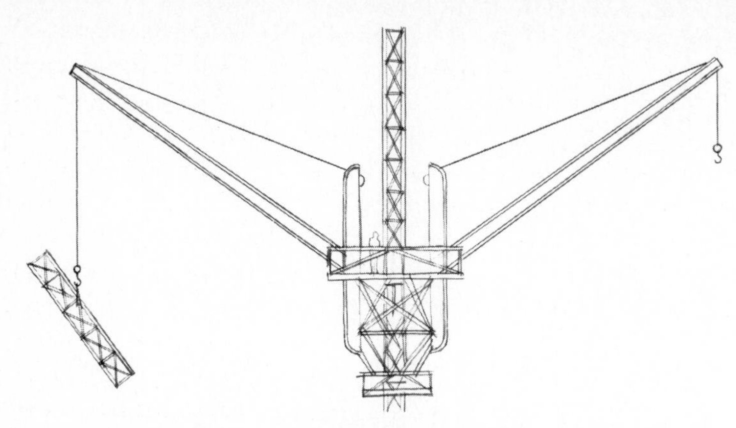

8.7 THE CENTRAL CREEPING CRANE

two temporary frames for its support. By this combination of cranes, winches, and creeping cranes the tower was erected in record time.

The vertical transportation of visitors was an altogether new and more difficult problem. The most efficient passenger elevators developed until then were the work of the American inventor Elisha Graves Otis, who moved the cabins by means of a piston sliding inside a cylinder through the introduction of water under pressure. His invention had two significant and new features: the elevators were hung from cables attached through pulleys to the end of the piston, so as to allow a large cabin-displacement with a piston of shorter length, and they had a safety device gripping the guides to prevent the fall of the elevators in the case of cable failure. European elevators at the time were instead directly supported from below by the vertical water-activated piston, and thus presented the safety feature of resting at all times on the piston. They could not fall under any circumstances. Eiffel, by contract, had to use French manufacturers for all parts of his Tower. He thus had no choice but to request bids from French elevator companies. But none of them were willing to bid on the elevators for the curved run between the first and second platforms, and Eiffel and the supervising committee had to accept the bid of a suspended elevator from Otis.

When all was said and done, the Tower found itself with three totally different vertical circulation systems. Two double-decker elevators ran inside the east and west buttresses from the first to the second platform. They were activated by an extremely complicated mechanism (never used again) consisting essentially of a gigantic bicycle chain moved by an enormous sprocket wheel (Fig. 8.8). Each Roux-Combaluzier elevator could carry 100 people at a speed of 200 feet per minute, making a tremendous racket, but achieving the required goal. (They were dismantled in 1900.) The elevators in the north and south buttresses were designed by Otis. They carried forty persons but ran smoothly at 400 feet per minute. One reached only the first platform; the other had a stop at the first, but continued to the second platform. Eiffel, the pragmatist, decided to test the safety of the hung elevator once and for all: he replaced the steel cables of the elevator with hemp rope, lifted the elevator fifteen feet and cut the ropes. The elevator slowly descended along the rails, braked by the Otis safety device, and proved its value.

The original proposal for the two upper elevators, going from the second platform to the top of the Tower, was extremely original in that it consisted of a central shaft shaped like a gigantic screw, along which the elevators would be moved up and down by rotating nuts driven by two independent electric motors, one for each elevator. The supervising committee, however, got cold feet at the thought of using electric motors and disapproved the scheme. Thus the two elevators are operated by the classical European system of hydraulic pistons, only thirty-six feet long, whose run is magnified by steel cables running over sheaves. All the elevator problems could have been easily solved by adopting a single piston-run system going all the way to the top of the Tower, but, for once, Eiffel put his foot down and refused to have the piston sunk 200 feet into the ground, exposed between the buttresses of the Tower, ruining its elegant, upward curving shape. Complicated as the adopted systems were, none interfered with the aesthetics of the Tower. If the mechanical systems do not have the simplicity and elegance of the

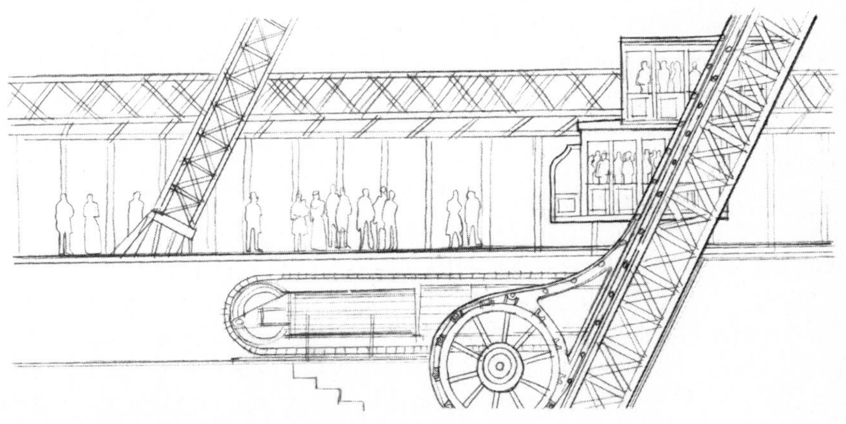

8.8 THE ROUX-COMBALUZIER ELEVATOR

structure, it must be said that they served their purpose and allowed visitors to reach the top of the Tower in about seven minutes. Ninety years have gone by since the inauguration of the Tower, and the original "speed" of its vertical circulation only hinted at the speeds to be reached by elevators in our high-rise buildings. It is, on the average, 2.4 feet per second in the Tower and 20 to 40 feet per second in our skyscrapers.

The Life of the Tower

Since 1889 tens of millions of visitors have been up the Tower, but not all have used its elevators. In 1906 a race was run up to the first platform and won by a man in three minutes flat. A slower time was clocked for a Mr. Dutrieux in 1959, but he did it on one leg. Going down, of course, is an easier task, and some one did on a bicycle in 1923 only to be beaten by another cyclist twenty-six years later, this time on a unicycle. If the main reason for climbing a mountain is the fact that "it is there," man-made mountains have the same attraction as real ones for mountain climbers and a 1,000-foot tower is bound to have for them an irresistible fascination. In 1954 a German student climbed to the top along the outside trusses of the Tower, stopping during the night for a short bivouac and reaching the top the morning after. But climbing was officially recognized as a proper and actually glorious way of reaching the Tower's top when a party led by a famous Italian mountain climber went all the way to the top along the west column, followed by millions of television viewers, to celebrate the seventy-fifth anniversary of the Tower's erection. And in true rock-climbing style, the leader rappelled from the top to the second platform along his rope.

Like other dramatic structures, the Tower has been since its opening a favorite jumping platform for suicides. An average of four people a year kill themselves from it, although the first did not jump but hanged himself from its beams. Only the Golden Gate Bridge can compete with the Tower in this lugubrious count: over twelve people a year have committed suicide from it since it was opened to traffic.

But not all that happened at the Tower was sad or crazy. Thanks to the unquenchable thirst of Eiffel for new scientific knowledge, during his lifetime it became an important station for the study of weather and, above all, of aerodynamics. Eiffel had an apartment in the Tower, still maintained as a museum, where he spent many hours dedicated to the new science of flight. He was the first to prove experimentally that the suction over the upper surface of a wing is more important to its weight-carrying capacity than the air pressure under it. His studies were recognized all over the world and particularly in the United States.

The gentleman divided his time between his studies, his investments, and his family, apparently unruffled by the storm of scandal in which he found himself involved. While he was designing and building the Tower—as if this committment were not enough for his energy—he had become the technical director of the new French company which, under the leadership of de Lesseps, the mastermind of the Suez Canal, was working on the first phase of the Panama Canal. Eiffel designed and began to build all the enormous steel gates for the canal and was directing the excavation through the jungle that had stymied all his predecessors. A combination of bribes, political corruption, anti-Semitism, and hostility from the United States put a sudden end to this gigantic endeavor and, for a while, tarnished the good name of Eiffel. Cleared of any malfeasance a few years later, he renounced any and all construction work and gave his time and attention exclusively to aerodynamics. The unflappable bourgeois gentleman did not seem hurt by the episode or by the renunciation of his lifetime career in construction. He wined and dined his friends in his apartment in the Tower and enjoyed his family, particularly on his birthdays, when every member had to be present to celebrate the unsmiling father and grandfather.

His autobiography, completed three months before his death, has not been published. (According to a statement by one of his grandchildren, he left only one typed copy of it to each of his five children.) Hence we do not know what Eiffel thought of himself, and it is most improbable that he would have given himself away even in a private work of this kind, but we have a glimpse of the man in a statement he is said to have often made: "I should be jealous of the Tower. It is much more famous than I am." Considering his indefatigable achievements through ninety years of life, we may understand his veiled regret. But if official Paris has been strangely cool towards her son, how many engineers have a monument erected, not in a cemetery but in the middle of a bustling metropolis, at once a center of life and a symbol of science and country, instantly recognizable by name the world over?

The Taller Towers

For forty years the Eiffel Tower was the tallest in the world. There are now four skyscrapers and at least four TV towers taller than Eiffel's masterpiece. A reinforced concrete tower in Moscow reaches 1,750 feet,

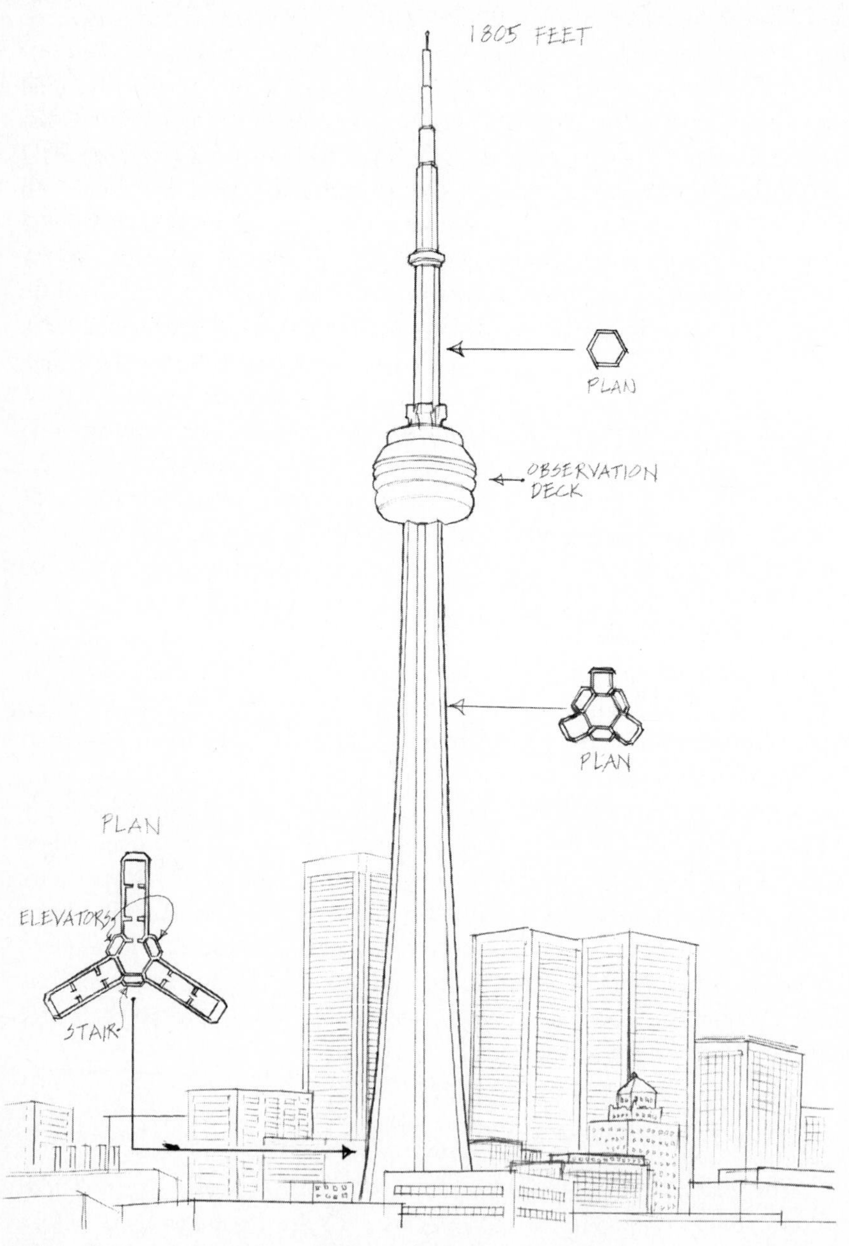

8.9 THE CANADIAN NATIONAL RAILROAD TOWER

the Canadian National Railroad Tower in Toronto is 1,805 feet tall, with the help of a 300 foot antenna (Fig. 8.9). The antenna in Blanchard, North Dakota is 2,063 feet tall, but it cheats: it is a guyed tower, restrained by inclined steel wires. And the same can be said of the 2,117 foot TV antenna in Warsaw.

Higher towers will certainly be built in the near future now that both structural theory and aerodynamics have acquired sounder knowledge. But they will never mean what the Eiffel Tower meant, first to the French and then to the world. They will be built for a purpose and not for an idea; they will attract attention but not become symbols; they will show how easy it is to improve on a first achievement once it becomes obvious that such an achievement is possible.

The Eiffel Tower will always be "the tallest tower in the world," since the achievements of man are measured by more than mere feet.

9 Bridges

Arches

The world of bridges is so infinitely varied that the layman must be somewhat baffled by it. Why is one bridge as light and elegant as a suspension span and another as massive as a medieval castle of brick, arches, and buttresses? Why does one steel bridge have an arch above the roadway and another an arch below it? Why do some bridges have an arched shape while others are straight? Why are some made out of reinforced concrete and others out of steel?

The answers to all these questions, and many others, rest on the fact that the type of bridge best suited to a particular site depends on a large number of factors. The length to be spanned, the nature of the river banks, the free height required under the bridge, the variations in the river's water level, the materials and the specialized labor available for its construction, the kind of traffic present and that to be expected, the kind of road approaches usable, and, last but not least, economic and even aesthetic considerations together with the preferences of the engineer—all play a role in the choice of a bridge type. If one considers all the possible combinations of these factors, then no large bridge can be identical with any other. Each is, or ought to be, the best structural solution to a very specific problem and, since bridges are almost all-structure, one should look at them first from the structural point of view to understand how the various types work. Their behavior is quite simple to grasp as the infinite world of bridges divides itself into at most four or five basic types. Let us look first at the arch bridge.

If a weight is hung from a piece of string held in two hands, the string takes a triangular shape with two straight sides meeting at the point where the weight hangs. If two weights hang, separately, from the string, its shape changes to one with three straight sides. When many weights hang from the string, the string shape has many short, straight sides and looks rather like a curve (Fig. 9.1). The cables of a suspension bridge are loaded by a large number of weights hanging from the numerous *suspenders*. If we could look at them with binoculars, we would see a shape consisting of many straight sides, but so many of them that from far away the cables look smoothly curved. The varying shapes taken by a string or a cable under varying weights are called *string* or (from the latin funiculum-string) *funicular polygons*, that is, many-angled figures.

Very thin, flexible elements can work only in tension. Strings and cables are so flexible that they cannot resist compression or bending, as a beam does. They can only resist pulls, and since they straighten when pulled, they are always straight between hanging loads. This is why, in order to carry loads by tension only, cables must change shape whenever loads change in location or number.

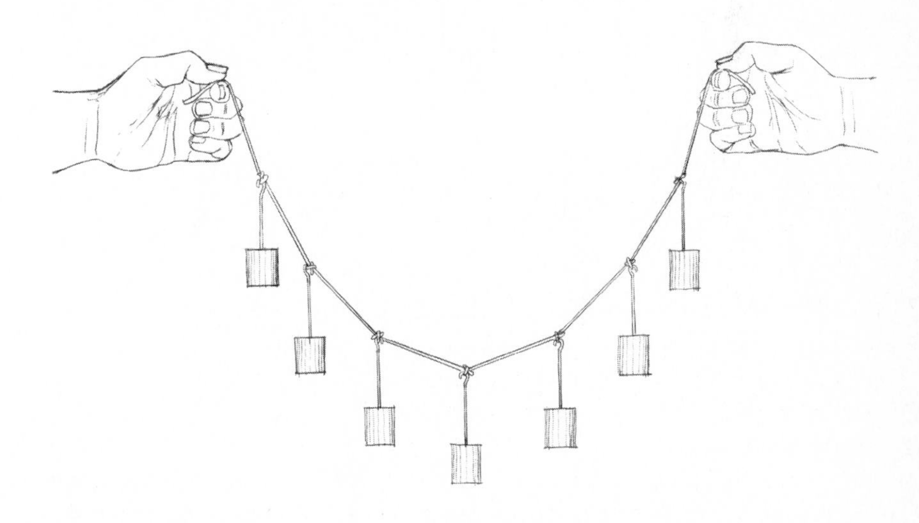

9.1 THE STRING POLYGON OF MANY WEIGHTS

The adaptability of cables to carrying different loads in tension, while wonderful, presents practical difficulties. Our structures are submitted to varying loads, but it would be inconvenient to have structures change shape all the time. Thus, to stop a cable from changing shape, we must stiffen it by means of a beam or truss, which we have seen in Chapter 5 to be rigid in bending.

In supporting a single weight by means of a string held in two hands, we may notice an important fact. Besides pulling up on the string ends to balance the weight pulling down, we must also pull *outward* to counteract the tendency of the two string ends to move towards each other (Fig. 9.2). These outward pulls of our hands are called *thrusts*. Similarly, in a suspension bridge the anchorages (massive masonry or concrete blocks built into the ground,) pull not only down, but outward on the cables to provide them with the needed thrusts.

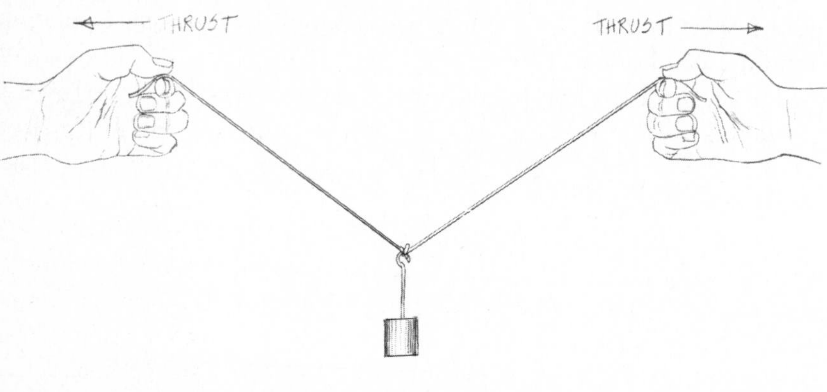

9.2 CABLE THRUSTS

Once the behavior of a cable is grasped, one easily realizes that an arch is nothing but an upside-down cable. Imagine flipping a loaded cable over after freezing it in its curved shape. The cable becomes an arch. The pull (or tension) in the cable becomes a push (or compression) in the arch, and the outward thrusts on the cable become the inward thrusts on the arch, which prevent it from opening up (Fig. 9.3). Of course, to freeze the arch shape, often called the *anti-funicular* of the loads, we must make it stiff, and hence much thicker than a slender cable, or it will buckle, as all thin, compressed elements tend to do. And because the arch has to be stiff to prevent its buckling, it does not need a stiffening

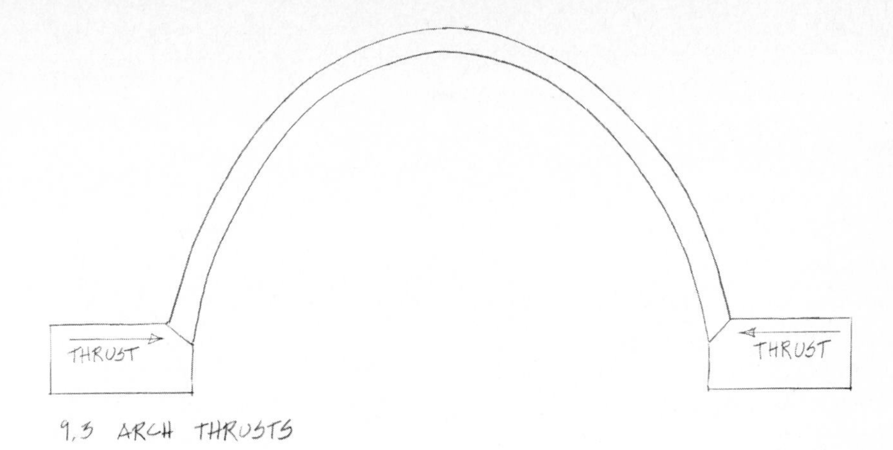

9.3 ARCH THRUSTS

truss! It keeps its shape under a variety of loads and is said to be stable, while a cable without a stiffening truss is unstable.

Since an arch is compressed all over—its own weight, the roadway, and the traffic loads all push down on it—it can be built of materials strong in compression, like stone, brick, and concrete. The availability of natural compressive materials explains why our ancestors built arch bridges and arched roofs over 2,500 years ago. The Romans, who were masters of masonry construction, used arches profusely and in a variety of applications. Masonry and brick arch bridges were found all along their road network and also in their monuments. The Coliseum's outer walls are pierced by arches, the Roman baths were covered by arched vaults, and the Pantheon has a large domed roof that works somewhat like a series of arches set around a circle (see Fig. 13.5). Their aqueducts carried water along the top of as many as three sets of superimposed arches. The maximum span of a Roman arch was about 100 feet and its shape was always that of a half-circle, because of the ease of erecting circular wooden scaffolds or *centerings* for their construction. While a long-span cable can be spun from tower to tower without any intermediate support (see Chapter 10), a masonry arch is built of separate blocks or *voussoirs* which must be temporarily supported on a centering, still usually made out of wood. The arch is started simultaneously from both ends of the centering and when the top block or *keystone* is wedged between the two adjoining blocks (Fig. 9.4), the centering can be lowered since each half-arch leans on the other, as Leonardo observed.

Once the arch is built, its ends must be prevented from spreading apart. The inward forces or thrusts needed to keep the arch from opening can be provided by the banks of the river when these consist of solid rock. Otherwise, heavy buttresses of masonry or concrete must be built

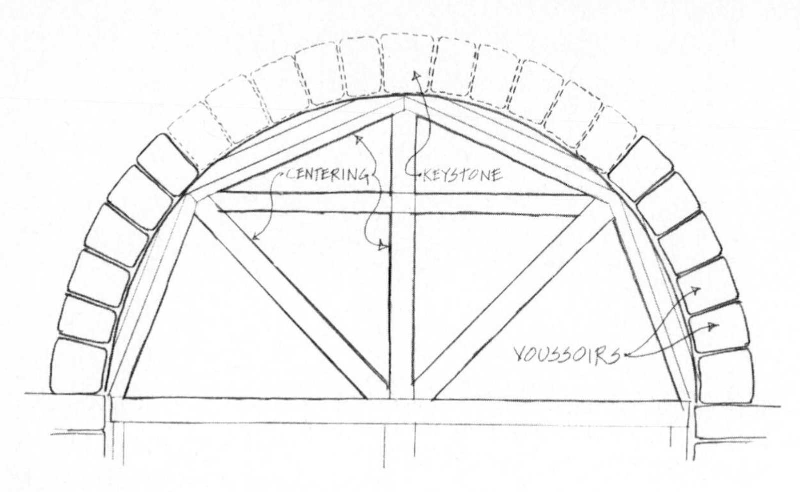

9.4 BUILDING AN ARCH

that resist the tendency of the arch to push outward, just as the anchorages resist the tendency of the suspension bridge cables to pull inward.

Modern arch bridges are often built of reinforced concrete poured in place or of large blocks (often hollowed to make them lighter) which are prefabricated on land and lifted on top of the centering, where their reinforcing bars are welded and the blocks are glued to one another by strong cement mortar. The longest reinforced concrete bridge built so far connects the Adriatic island of Krk to the mainland of Yugoslavia and spans 1,280 feet (Fig. 9.5). To carry the roadway, vertical columns or *struts* of varying length must be built up from the arch. (These are the counterpart of the suspenders in the suspension bridge.) The weight of the roadway and the vehicles pushes down on the struts that are compressed, while the roadway is supported by a series of beams spanning from strut to strut. These beams, stiff in bending, and the roadway itself contribute to the overall stiffness of the arch bridge and act very much like the stiffening truss of a suspension bridge. The extremely elegant bridges of the Swiss engineer Robert Maillart consist of very thin concrete arches (sometimes only seven inches thick over a span of 300 feet), which share the task of carrying the loads with the concrete beams and slabs of the roadway. An upside-down picture of a Maillart bridge looks exactly like a suspension bridge, explaining visually their opposite, but analogous behavior (Fig. 9.6).

9.5 KRK ARCH BRIDGE IN YUGOSLAVIA

The longest arch bridges in the world today are built of steel. They look and work very much like reinforced concrete bridges and usually have the advantage of being lighter because of the high strength of steel in compression. Now (1980) the very longest is the New River Gorge Bridge in West Virginia, which spans 1,700 feet (Fig. 9.7). Previously it was the Kill van Kull Bridge, connecting Staten Island to New Jersey, but not by much: it was purposely made just five feet longer than the famous steel bridge at the entrance of Sidney harbor in Australia so that the United States could boast the longest steel arch bridge in the world.

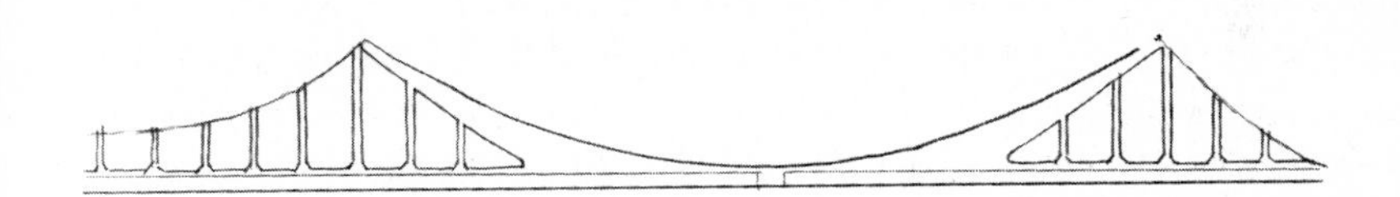

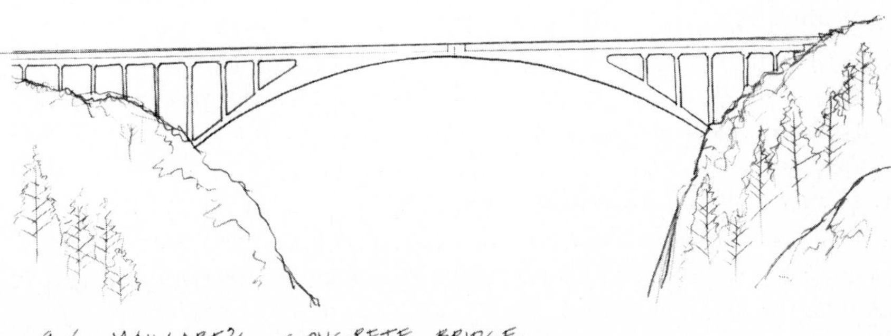

9.6 MAILLART'S CONCRETE BRIDGE

9.7 THE NEW RIVER GORGE BRIDGE IN WEST VIRGINIA

Railroad Bridges

The heroic age of bridge construction coincided with the expansion of the railroads in Great Britain and the United States. The first modern locomotive, the Rocket, was invented and built by George Stephenson in 1829, and the first passenger railroad was inaugurated in England in 1825. But the greatest of railway achievements was the spanning of the American West in the nineteenth century—the networks of the Baltimore and Ohio, the Erie, and the Pennsylvania railroads and the first transcontinental railway, the Union Pacific, completed in 1860.

Forgetful of the violence and chicanery which characterized this era, we have developed a great nostalgia for the old railroads, manifested in the expensive and widely appreciated hobby of toy trains. Both children and parents actively participate in it and while the first outgrow it, the last often don't. To carry the unprecedented loads of the locomotives and the railroad cars across deep gorges and wide rivers, all kinds of new bridges had to be built, with great urgency and increased strength. After a few wooden spans were erected, mostly in the form of *trestles* (Fig. 9.8) with their high, narrowly spaced towers of timber connected by short, heavy timber beams, the miraculous new iron material, steel, was universally adopted. Steel arches, large trusses, cantilevered bridges, lift, bascule and pivoted bridges mushroomed throughout the country. Meanwhile, after the example of England, France had built by 1840 the largest railway system in Europe. Germany, Italy, and Russia followed suit. The railroad gauge had been standardized to a width of four feet eight and one-half inches by 1880, and for about 100 years the railroad reigned supreme all over the world.

9.8 WOODEN TRESTLE

Possibly the most common structure used in railroad construction to span rivers and valleys was the *truss*. Hence, a brief discussion of railroad bridges must start with an analysis of this most ingenious extension of the beam, involving straight bars of steel working only in tension or in compression. As the span of a beam and the loads on it increase, its dead load increases rapidly out of proportion with the loads it has to carry. We have seen in Chapter 5 how efficiency is improved by giving the beam an I-shape, provided its depth is increased and the flanges are set farther and farther away from each other. Consequently, the web connecting them becomes deeper and heavier. And here is where the basic idea of the truss comes in. To lighten the web why not open up holes in it, obtaining a perforated I-beam? Finally, as the dimensions of the beam make its fabrication as a single rolled element impractical or impossible, why not build it up by connecting the upper bar to a lower bar by means of inclined and vertical bars (Fig. 9.9) which constitute, so to say, a discontinuous web? To make this built-up beam rigid its *meshes* or holes, created by the web bars, should be of a triangular shape, since the triangle as the great bridge builder John Roebling put it, is "the most indeformable geometric figure." A *triangulated truss* is thus obtained which can be made as deep as necessary and is substantially lighter than a full-web beam.

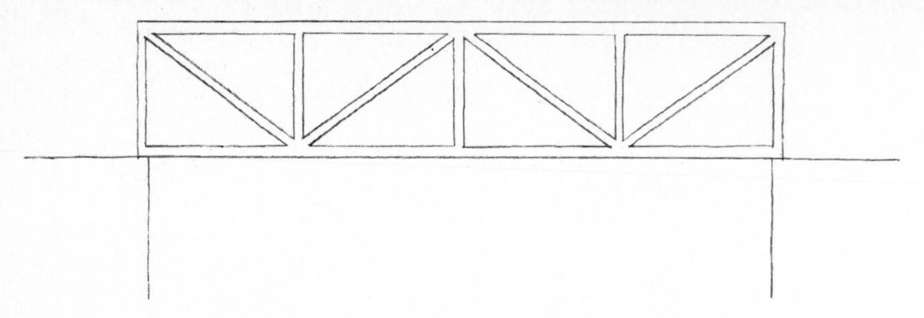

9.9 TRUSS BRIDGE

Since a truss supported at its ends is but a perforated I-beam, its upper flange or *chord* is compressed, while its lower chord works in tension (see Chapter 5). The web bars resist the equivalent tension and compression of the beam shear and work in tension or compression depending on their inclination. Since all the bars of the truss are either tensed or compressed, they work much more efficiently than the layers of a beam, in which the material near the middle, neutral axis does very little work at the expense of the material of the flanges. It is no wonder that trusses are used for 1001 applications in all kinds of structures, from those stiffening the core of high-rise buildings (which were called X-ed frames in Chapter 7) to the prefabricated trusses called *bar-joists* available on the market in lengths from 20 to 100 feet [with chords and diagonals of steel (Fig. 9.10a) or chords of wood and diagonals of steel pipe (Fig. 9.10b)], from the enormous steel trusses used in bridge construction to those made of laminated wood and used in roof construction.

Steel trusses with horizontal parallel chords were patented during the railroad era with a variety of patterns for the web bars (Fig. 9.11) and each is called to this day by the name of its inventor. But soon the concept of triangulating a steel structure was extended to trusses with one or both chords in a curved shape, and finally large steel bridges were built as triangulated arches, as they are to this day (see Fig. 9.7).

Prior to the advent of the railroad most of the freight traffic, of course, travelled through a network of canals and rivers on barges, pulled first by mules and eventually by locomotives. Memories of our canals still linger in the lovely songs of the nineteenth century and the words "Erie Canal" evoke dreams of peaceful travel on slow water through the locks. The canal lobby fought the railroads. Whenever a railroad bridge was planned to cross a river or canal at grade level, a cry arose about impeding the water traffic, and the bridge was required to cross high above the water.

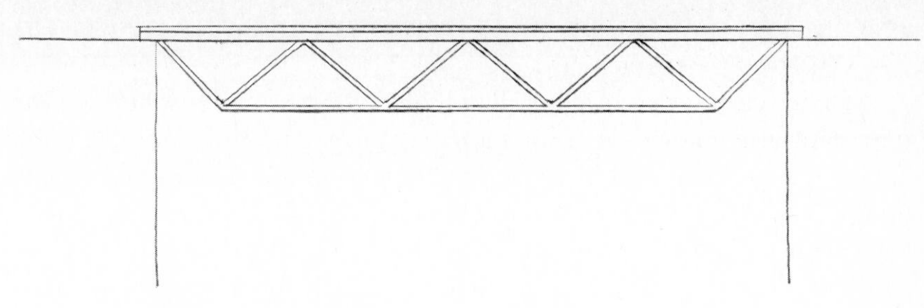

9.10a STEEL BAR JOIST

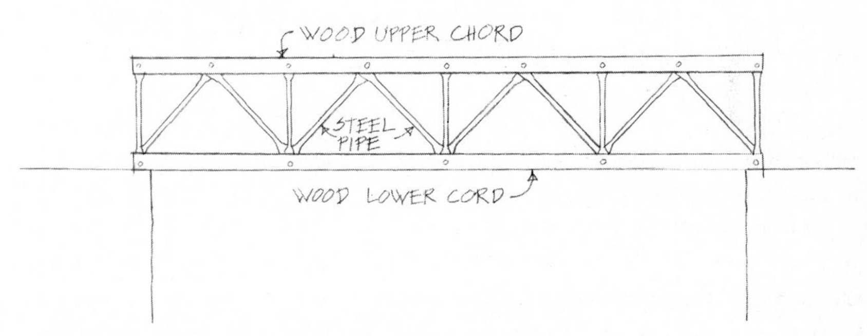

9.10b BAR JOIST WITH WOOD CHORDS

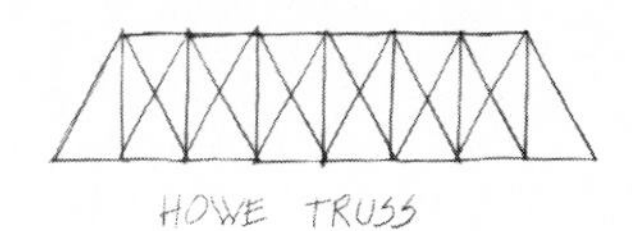

HOWE TRUSS

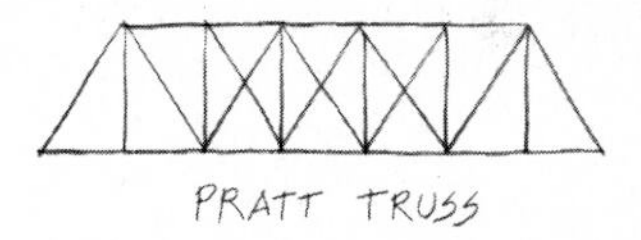

PRATT TRUSS

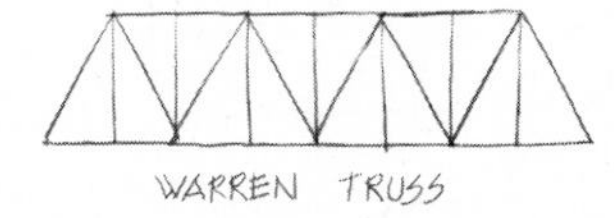

WARREN TRUSS

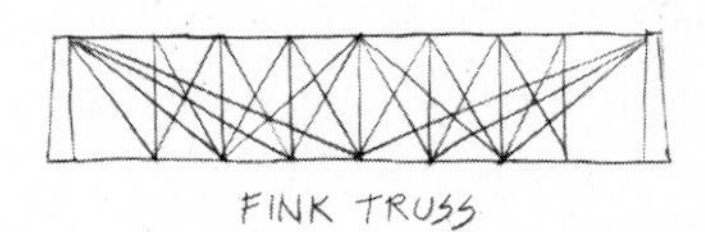

FINK TRUSS

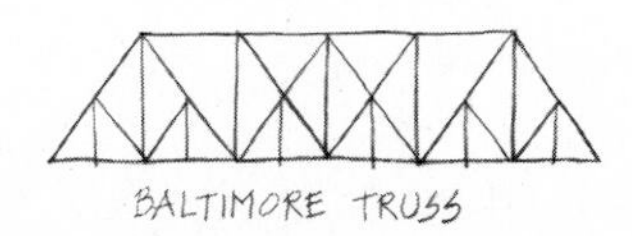

BALTIMORE TRUSS

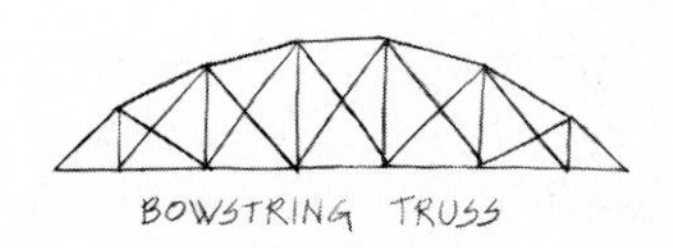

BOWSTRING TRUSS

9.11 PATENTED STEEL TRUSSES

These clearances reached hundreds of feet under some of our important bridges, and such rules are even today strictly enforced by the Corps of Engineers of the U.S. Army, who have jurisdiction over water traffic. This makes the erection of a bridge at times a very costly enterprise. One of the reasons for the variety of bridge types lies in the effort to beat this problem. To begin with, the roadway of an arch bridge can be located at the level of its supports or *abutments,* whenever these are high enough above the river. Two birds are killed with one stone in this case, since the roadway and the approaches to the bridge are cheaper and, moreover, the roadway acts as a tie between the arch ends, preventing their spreading out, and supplying the needed thrusts. When the bridge crosses a deep gorge, an "inverted arch" can be set under the roadway, really acting as a tensed cable, with the advantage of allowing the use of smaller bars, since no additional strength is required against buckling in compression.

If the roadway cannot be located in its lowest position, it may be built half-way up the arch or above it. In its lowest position the roadway hangs from the arch; in its highest it is supported on it by means of compressive struts; in the intermediate location it is partly hung and partly supported.

9.12 CONSTRUCTION OF EADS'S MISSISSIPPI BRIDGE

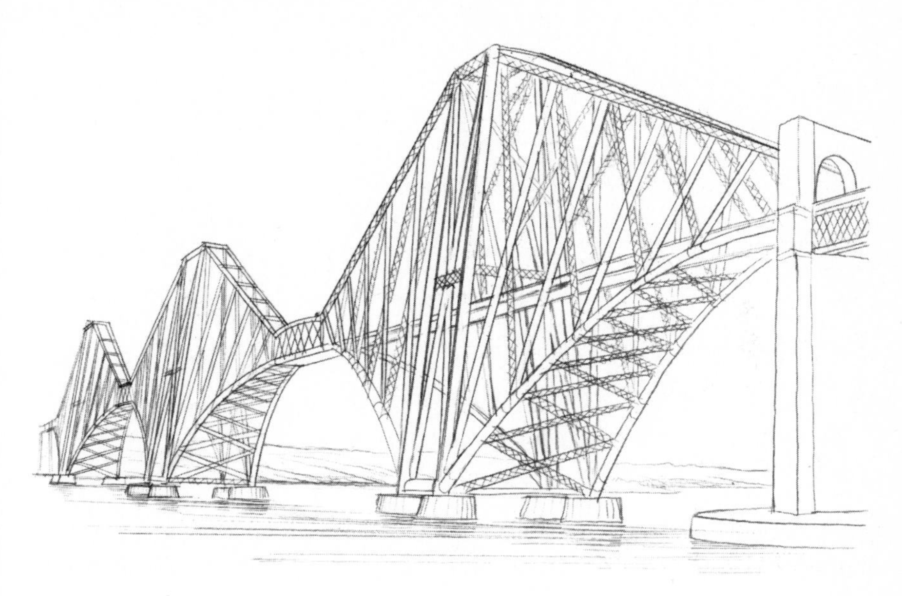

9.13 FIRTH OF FORTH BRIDGE

The next ingenious idea is aimed at allowing a free flow of river traffic during construction of the arch by eliminating its centering founded at the bottom of the river. It was Eads, the great bridge engineer of the second half of the nineteenth century, who first conceived of erecting the two halves of the arch as cantilevers by supporting them at their tips with cables from temporary towers until the two cantilevered halves met and were joined at midspan. The great three-arch steel bridge over the Mississippi was built by Eads by this daring new method in 1867 (Fig. 9.12). It was almost a natural step to build next cantilevered arches in which the temporary cables were replaced by permanent steel members, thus obtaining a bridge supported both by its upper members acting as cables and its lower members acting as arches. One of the greatest bridges ever built, the Firth of Forth Bridge in Scotland with a span of 1700 feet, cantilevers its three sections from three piers but has in addition two truss bridges end-supported on the tips of the cantilevers (Fig. 9.13). Although erected in 1890, it remains the second longest bridge of its kind in the world and was overtaken by the Quebec Bridge only in 1917.

9.14 BASCULE BRIDGE

The last expedient to free the navigable channel from impediments during the arch (or truss) construction consists in building the arch on its centering away from the site, then floating both centering and arch to the site, and finally rapidly connecting the arch to its abutments. This procedure is also most efficient when the waters under the bridge are subject to short, violent floods or storms. The greatest of French concrete engineers, Freyssinet, used this method to erect the monumental Plougastel Bridge in northern France in 1920, lifting the heavy arch from the centering by means of hydraulic jacks.

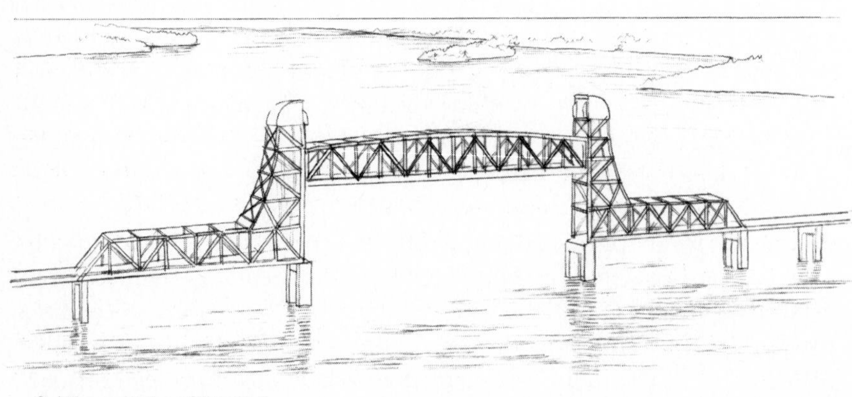

9.15 LIFT BRIDGE

To get a low bridge out of the way when river traffic must go through, a variety of movable bridges has been invented. In one type the center span consists of two trusses, each pivoted at one end and capable of rotating upward when a boat needs to go by. It is called a *bascule bridge* (Fig. 9.14). Or the center span may be lifted between two end-towers, as in so many bridges in the New Jersey industrial area and the large bridge over the Harlem River in New York, the first link of the Triborough Bridge system (Fig. 9.15). Or finally the entire center span of a bridge may be made to rotate horizontally on a central pier, as in some of the bridges crossing the Harlem River, thus clearing two separate channels on the sides of the pier (Fig. 9.16).

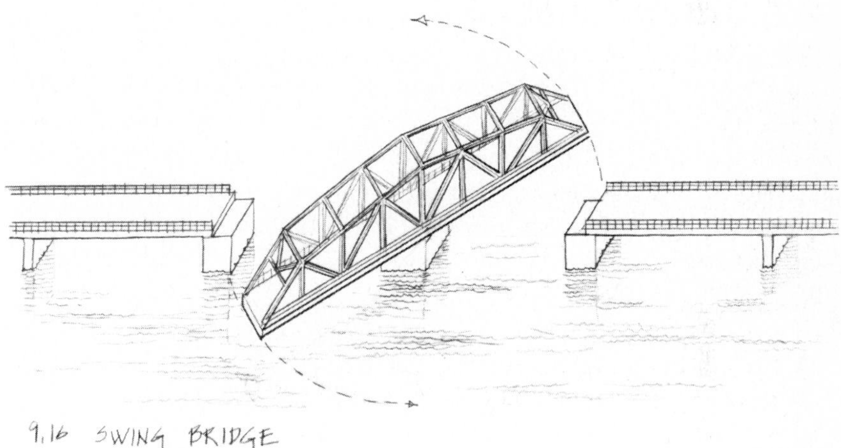

9.16 SWING BRIDGE

Concrete Bridges

Some if not all of the bridge concepts born under the impact of the railroads and executed in steel have been extended to concrete. The trestle, built of wood at first and then of steel, has acquired a new elegant expression in the long viaducts built in Europe. These consist of hollow piers of reinforced concrete, at times 200 to 300 feet high, over which runs a roadway of hollow reinforced or prestressed concrete pipes of rectangular shape which are prefabricated on the river banks and slid into position. In the latest application of this principle the roadway pipes are extruded continuously, like spaghetti from a pasta maker, from both ends of the bridge until they meet at midspan. The simplicity, slenderness, and pure geometric shape of these viaducts make them works of beauty (Fig. 9.17), showing that high technology does not have to sacrifice aesthetics to economy.

One of the longest bridges in the world, over five miles long—the Maracaibo Bridge in Venezuela (Fig. 9.18) designed by Riccardo Morandi—is built in concrete with a complex sequence of twenty end-supported spans of 120 feet, followed by seventy-seven spans of 150 feet (cantilevered from trestles and supporting shorter spans on their tips), and by five 780-foot cantilevered spans supported by cables at their ends, which in turn support short spans on their tips on the same principle as the Firth of Forth Bridge. The construction of this monumental work took only forty months, and its roadway has a clearance of 150 feet over the navigable channel.

Longer bridge and tunnel systems are used to cross wide stretches of water, like the Chesapeake Bay Bridge, consisting of twelve miles of concrete trestles and including a two-mile long tunnel. But the longest bridges in the world are those that do not have to contend with water traffic and can be laid right on the water. These *floating bridges* consist of reinforced concrete hollow barges or *pontoons* anchored to the bottom of the water by draping steel cables (Fig. 9.19) and connected to each other by elastic joints that allow them to move slightly with respect to

9.17 CONCRETE VIADUCT

9.18 MARACAIBO BRIDGE IN VENEZUELA

each other in case of violent storms. If water traffic is necessary, one of the pontoons rotates on a vertical pivot at one end and opens up a gate while floating in the water. The Hood Canal Pontoon Bridge over Lake Washington in Seattle is 6,250 feet long.

If the magnificence of modern bridges cannot be appreciated from a fast moving automobile, there are few experiences to equal crossing a large bridge on foot. The half-mile walk over the red Golden Gate Bridge in San Francisco, suspended over the blue waters of the Pacific and profiled against the cloud-dotted sky, the short promenade over the middle deck of the Brooklyn Bridge against the background of the New York skyline, the leisurely bicycle ride over the George Washington Bridge with the view of Manhattan from the high-rises of the Columbia Medical Center to the towers of the World Trade Center, are as exciting as the passage along the heavy cantilevered trusses of the Queensborough Bridge or that on the Williamsburg Bridge reached through the cavernous stairs in its towers. And what is more exhilarating than to cross the suspension bridge over the Tagus River in Lisbon, surrounded by a landscape of incredible beauty and the memory of Columbus sailing for the Indies, or to walk the Bosphorus Bridge connecting Europe to Asia? Human enjoyment rests on physical contact with reality. Bridges can best be

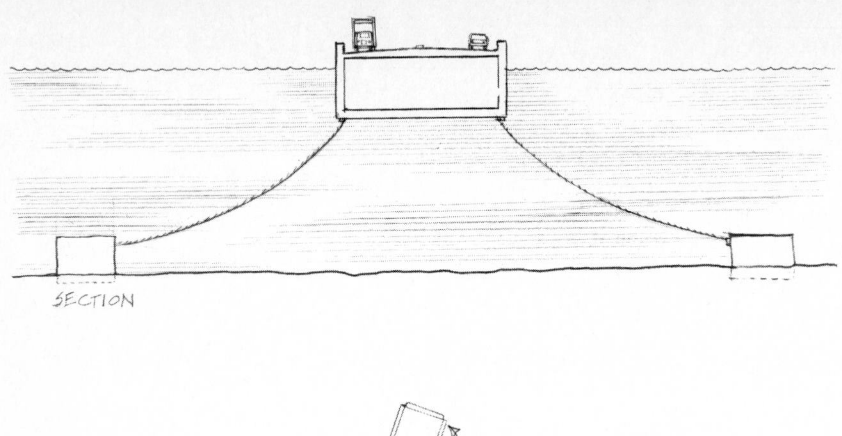

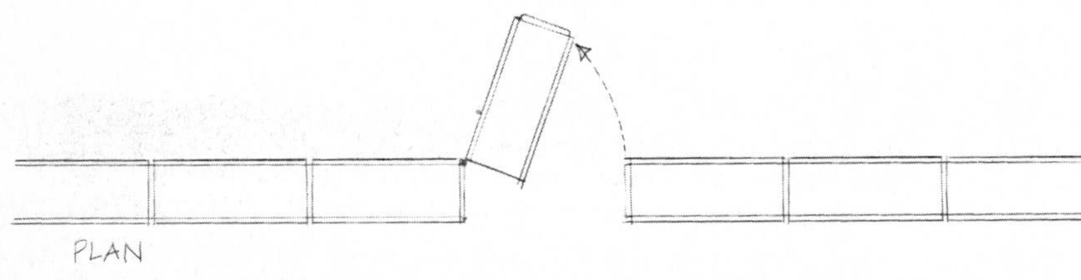

9.19 FLOATING PONTOON BRIDGE

understood and appreciated as expressions of the human spirit and works of subtle beauty when crossed on foot. Humanly useful technology can, in fact must, go hand in hand with beauty. The hideousness of some of our surroundings is not inherent in the development of our technical culture.

Space Frames

A large horizontal roof over a rectangular area, unimpeded by intermediate columns, is but a bridge in two dimensions, which can be crossed from any of the walls supporting it to its opposite wall. Some of the bridge systems in steel and concrete have, therefore, been adapted to fill the roofing needs of our large convention and exhibition halls.

The inventiveness of Pier Luigi Nervi suggested the use of a suspension bridge for the roofing of a paper plant in Mantua, Italy, which demanded a clear span of 830 feet with a width of 100 feet. The roof towers were inclined backwards to resist the thrusts of the cables and were propped by shorter concrete struts so as to look like gigantic Greek letters lambda, while the roof surface was the "deck" of the suspension bridge (Fig. 9.20). In collaboration with the steel engineer Covre, Nervi thus produced one of the longest roofs in the world with great economy and, as usual in his work, pleasing aesthetic results.

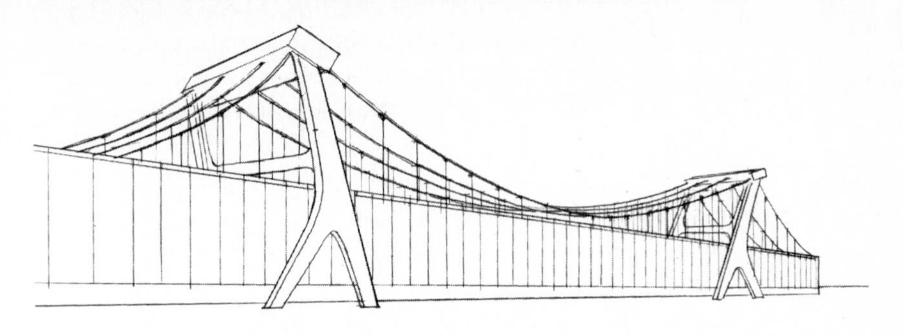

9.20 BORGO PAPER PLANT IN MANTUA, ITALY

The same principle of the suspension bridge, but in a cantilevered form using essentially one-half of a bridge, has been used in airplane hangar roofs, which require one entire openable wall to allow the entrance and exit of the planes. Such cantilevered roofs can be seen in some of the hangars at the J.F. Kennedy Airport in New York and in other airports around the world. At JFK a cantilevered cable-supported roof has been arranged in an elliptical pattern for the Pan Am terminal with cantilevers of 150 feet (Fig. 9.21).

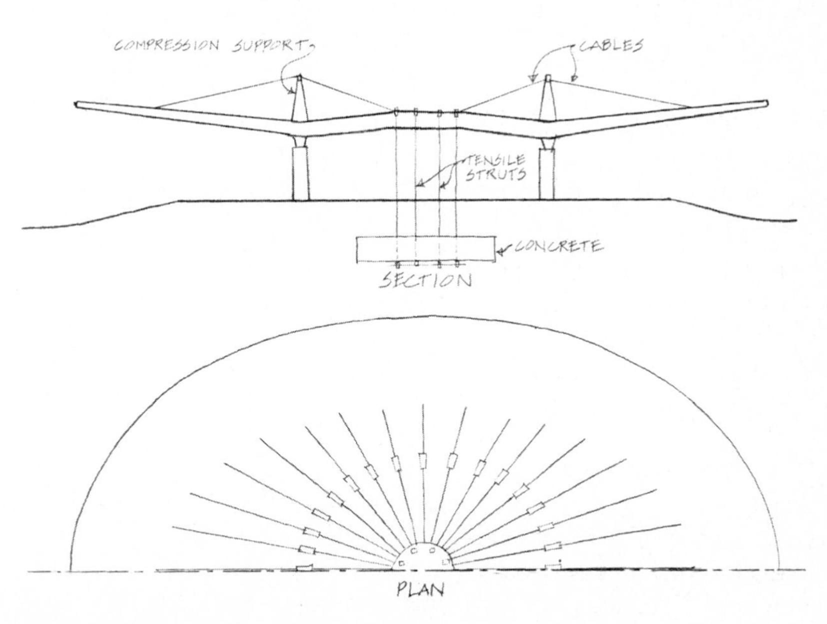

9.21 THE PAN AM TERMINAL AT JFK AIRPORT IN NEW YORK CITY

For more nearly square areas, like those of convention and exhibition halls, it would seem logical to support the roof surface on a rectangular grid of vertical trusses which can be built of bars of few different sizes. Such systems would be practical but for two reasons: the cost of the joints formed by the trusses meeting at right angles and, more importantly, the fact that the system consists of rectangles, and rectangles are not inherently stiff. One of the great minds in the field of American invention, Alexander Graham Bell, was the first to realize that if such a roof could be triangulated in space, it would acquire a much greater stiffness in all directions and hence could be made lighter. Thus the modern *space frame* sprang from the mind of an electrical engineer and gave rise to a whole family of roofs having the enormous advantage of modular construction, easy assemblage, economy, and visual impact. These roofs consist essentially of a number of pyramids, some with bases up and others down, creating two parallel horizontal grids of bars interconnected by zigzagging bars in at least three directions (Fig. 9.22). One obstacle postponed their adoption: the mathematical calculations required by their design are so time-consuming that they did not become practical until the advent of the electronic computer. Thus one field of engineering had to rely on another before a new conception could become reality.

9.22 SPACE FRAME

The space frame is today one of the most economical structures for covering large areas. The largest was erected in Brazil, where the translucent plastic roof of the exhibition hall at Anhembi Park in São Paulo is supported by an aluminum space frame covering an area 853 feet square, subdivided by tubular tripod supports into squares 197 feet on each side. Such is the visual appeal of the zigzagging bars of a space frame

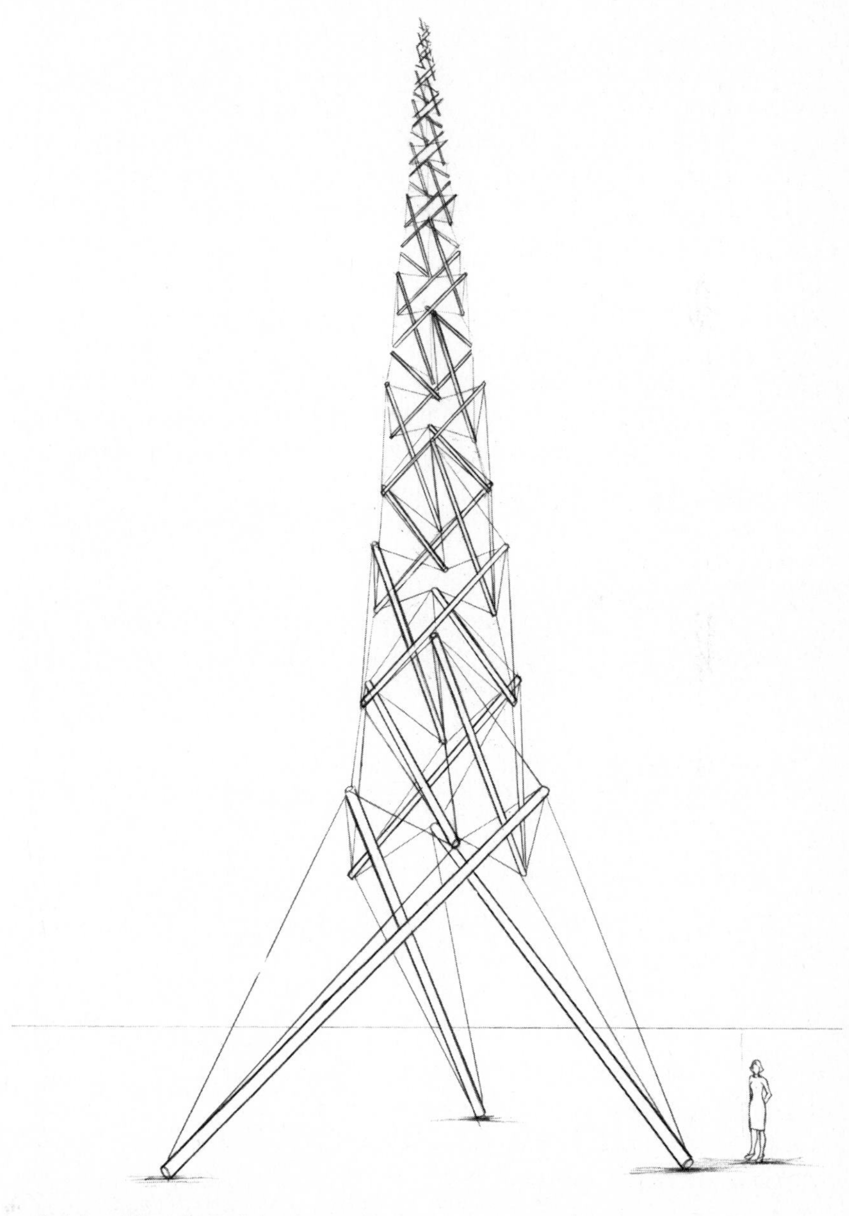

9.23 TENSEGRITY SCULPTURE BY KENNETH SNELSON

that they were used (by I.M. Pei) in the roofing of the magnificent main hall of the new wing of the National Gallery in Washington as well as in the roofing of the immense Air and Space Museum in the same city. And just for the sake of their appearance, vertical space frames have been used as transparent walls on entire sides of large buildings or as roofs over interior courtyards. The structure of both the walls and the roof of the new Convention Hall in New York City, occupying an area of two by five city blocks, consists of a single space frame covered with reflecting glass.

It is hard to put into words the many reasons for the visual appeal of space frames: their lightness, their transparency, and their geometry, which seems to vary dramatically with a change in vantage point, certainly contribute to it. But their aesthetic content is so high that Kenneth Snelson has become famous the world over for his beautiful Tensegrity sculptures, which are nothing but space-frame towers and girders floating in space. Tensegrity, invented and patented by Snelson, has added a new component to the elegance and airiness of the space frame. Since its bars work either in compression or in tension, Snelson uses aluminum or steel pipes as compressed members and connects them with a continuous steel cable that constitutes the tension bars. He does this in such a way that no two pipes touch each other (Fig. 9.23). Thus, no continuous compression path can be discovered in the structure of his sculptures, and the compressed pipes seem to float independently in space.

Down-to-earth engineers of the nineteenth century invented the humble truss for purely utilitarian purposes; now it has become an expression of art. This may be a lesson to the detractors of technology, who emphasize its negative aspects and ignore the many faceted meanings of so many of its fruits.

10 The Brooklyn Bridge

The Creator of the Bridge

On May 24, 1883, the mayors of Brooklyn and New York, Low and Edson, Governor Cleveland of New York and President Chester Arthur, followed by a throng of thousands, inaugurated the Brooklyn Bridge (Fig. 10.1). The engineer in charge of its construction, Washington Roebling, was not there. Paralyzed from the waist down by the "bends" he suffered while working in the underwater caissons for the towers' foundations, he watched through binoculars from his house on the Brooklyn shore. What he saw was a masterpiece conceived by John Roebling, his father, thirty-six years earlier and built by the son over a period of fourteen years.

The Brooklyn Bridge was a reality and a symbol: the longest suspension bridge in the world with a main span of 1595.5 feet, a monument to the persistence of two men and a woman, the two Roeblings and Emily Warren, Washington's wife. It is, perhaps, the most beautiful bridge in the world. There is no view more exciting than that of the Manhattan skyline at sunset seen through the network of inclined stays, vertical suspenders, and curved cables of the bridge. It made Greater New York a possibility by uniting its two most populous boroughs. It was conceived so wisely that, notwithstanding the unexpected increase in traffic, it needed strengthening only by the addition of light trusses, seventy long years after its construction. Designated a landmark in 1964, it carries a

10.1 THE BROOKLYN BRIDGE

plaque on the west face of the Brooklyn tower which reads:

THE BUILDERS OF THE BRIDGE
DEDICATED TO THE MEMORY OF
EMILY WARREN ROEBLING
1843–1903
WHOSE FAITH AND COURAGE HELPED HER STRICKEN HUSBAND
COL. WASHINGTON A. ROEBLING, C.E.
1837–1926
COMPLETE THE CONSTRUCTION OF THE BRIDGE
FROM THE PLANS OF HIS FATHER
JOHN A. ROEBLING, C.E.
1806–1869
WHO GAVE HIS LIFE TO THE BRIDGE
BACK OF EVERY GREAT WORK WE CAN FIND
THE SELF SACRIFICING DEVOTION OF A WOMAN
THIS TABLET ERECTED 1951 BY
THE BROOKLYN ENGINEERS CLUB
WITH FUNDS RAISED BY POPULAR DEMAND

John A. Roebling, the designer of the Bridge, was born in Germany and studied architecture and engineering in Berlin. A liberal opposed in principle to the Prussian government, he was inspired to come to the United States by no less a teacher than the philosopher August Hegel.

Upon arriving in America, Roebling and his young friends established the town of Saxonburg in Pennsylvania on a plan drawn by Roebling, who soon after conceived an idea that made himself and Saxonburg rich. He suggested the substitution of steel wire rope for the hemp hawsers then used to pull the boats of the Portage Railroad. Roebling and his friends built a factory for the wire's manufacture. The first step towards the erection of the Bridge had been taken, and the Roebling Wire Rope Company is to this day one of the most important manufacturers of steel cables in the world.

In 1841 the thirty-five-year-old Roebling got his first chance to use steel wires in a suspension bridge, over the Allegheny River. This led to his second stroke of genius: he proposed not only to reconstruct an aqueduct over the river as a suspension bridge, but to build its cables using a new procedure. Rather than lay all the wires constituting a cable on the ground, as was traditional, then fastening them together and lifting them up to the top of the towers, Roebling proposed to spin the wires from one anchorage in the ground, over the towers, to the other anchorage and to tie them together in their naturally acquired curved configuration. This method required fording first a few light ropes of twisted steel wires across the river, lifting them up to the towers by means of hoists, and then spinning the cable wires back and forth from one anchorage to the other by shuttling along a catwalk hanging from the ropes. His method is still used in all suspension bridges today, when the weight of the cables has become so great that it would be impossible to lift them to the top of the towers after they have been assembled.

On May 17, 1854, the Wheeling Bridge over the Ohio River, spanning 1010 feet and designed by Roebling's rival Charles Ellet, collapsed during a wind storm. A local reporter described the disaster: "Lunging like a ship in a storm, the deck rose to nearly the height of the towers, then fell, and twisted and writhed, and was dashed almost bottom upward." (By substituting Tacoma for Wheeling this becomes the exact description of the Tacoma Narrows Bridge failure in 1940 due to aerodynamic effects.) Roebling, undaunted—it had not been his bridge—rushed to propose a new suspension bridge on the Ohio, and his superb engineering intuition led him to a third stroke of genius: inclined cables or *stays,* connecting the tops of the towers to various points on the sides of the deck. These prevent the opposite up-and-down motions of the deck sides and stop any twisting aerodynamic oscillations. Again, Roebling's stays are used in some modern suspension bridges, and this concept has been recently extended to the idea of supporting the deck by stays alone, doing away with curved suspension cables. Pioneered by German engineers, *stayed bridges* are rapidly taking the place of suspension bridges for spans of intermediate length (Fig. 10.2).

These were ingenious and creative times. While Roebling was rebuilding the Wheeling Bridge, his rival Ellet was busy with the construction of a suspension bridge over Niagara Falls (this bridge eventually was finished by Roebling). Obviously one could not ford the initial ropes through the onrushing waters, but it occurred to Ellet to call for a competition among youngsters, with a prize of $5 to the first who would fly a kite across the falls. The competition was successful, and Ellet attached to the kite string a very light steel wire, then a heavier one to the light wire and continued the process until his first wire bundle exhibited its lovely catenary shape across the falls. Ellet's idea has been extended to the erection of large "balloon roofs" by carrying the steel cables required by this type of roof (see Chapter 15) from one anchorage to the other by means of a helicopter.

As if to rehearse the great feat of the Brooklyn Bridge, John Roebling was entrusted in 1856 with the construction of a 1,200-foot span over the Ohio River in Cincinnati. His bridge looks almost exactly like The Bridge, except for some minor stays attached under, rather than above, the deck. Roebling was almost sixty and, incredible but true, ready then to "leave bridge-building to younger folks." His wish was not to be satisfied.

10.2 STAYED BRIDGE

The thought of connecting Brooklyn to Manhattan had occurred often since the year 1800 and had always been rejected as a physical or financial impossibility. But in 1866 a company was chartered by New York State to build a totally undefined bridge between the two cities, and Boss Tweed of Tammany Hall, the Democratic New York stronghold, and a number of his cronies "bought" stock in the company, at an eighty percent "discount" after a first bribe—the first of May—of $65,000. Like the Roman emperor Vespasianus, who set a tax on public urinals and answered the objections of his advisers by the famous saying: "*Non olet*" (money never stinks), Roebling, a man of absolute integrity, accepted this shady participation, well-knowing that without the help of the politicians his dream would never become reality. The project, almost complete in every structural and construction detail was approved by an advisory committee of the best civil engineers in the country three years later. When the corruption of the Tammany Ring was exposed, Tweed and his friends were expelled from the charter company. By 1876 the federal government required the mayors, comptrollers, and chief aldermen of the two cities to sit on the executive committee for the bridge, and some kind of honesty was restored to its administration.

The creator of the bridge was barely able to start its construction. The year of the bridge approval, after he had sent his son Washington and his wife Emily to Europe for almost a year to study the use of underwater air-pressured caissons for the foundation of bridges, John Roebling's right foot was caught between the boat fender and the pilings of a ferry slip from which he was studying the Brooklyn tower location. Tetanus developed, and on July 21, 1869, John Roebling died.

A month later his son, Colonel Washington Roebling, a graduate of Rensselaer Polytechnic Institute and a hero of the Civil War, was appointed to his father's position as Chief Engineer for the bridge. The saga of its construction had started.

The Construction of the Bridge

The first step in the construction of a bridge, built on piers or towers under the level of the water, is the placement of their foundations at the bottom of the river. James Eads, one of the great bridge engineers of the nineteenth century, had learned the technique of compressed-air caissons for bridge foundations during a convalescent trip to France. The caissons consist of enormous water-tight boxes or *working chambers* (Fig. 10.3) that have no bottom, from whose top a well-like steel shaft extends above the water level. A pump pushes compressed air through pipes in the shaft into the working chamber until the water in the chamber is pushed out. The workers, soon nicknamed *sand hogs*, descend into the submerged chamber and, breathing the highly compressed air, excavate the soil at its open bottom. Other workers meanwhile build the bridge towers of brick and stone on the solid top of the working chamber. As the soil is excavated, the cutting steel edges of the chamber's sides penetrate into the soil, pushed down by the increasing weight of the tower being built on its roof. This operation is continued until the cutting edges reach solid rock or sufficiently strong soil, at which time the chamber and the steel shafts are filled with concrete and constitute the foundation of the tower, which, partly under water, is built above water to the height required by the design. When a sand hog has to move down to the working chamber, he first stops in a small *decompression chamber* or *air-lock* at the top of the shaft (Fig. 10.3). This is closed tight and pumped full of air until it is at the same pressure as the working chamber. The sand hog can then proceed down by opening a hatch at the bottom of the air-lock, whose pressure has been equalized to that in the working chamber. On the way out the sand hog must again enter the air-lock through a bottom hatch and wait until the pressure in it is decreased and equalized to that of the outside air. The hatch in the ceiling of the air-lock is then opened and the sand hog can get out. This last operation is critical. It must occur very slowly because the nitrogen gas which at high pressure is dissolved in the blood, forms bubbles if the pressure is suddenly or rapidly reduced, and these bubbles can impede or even stop the blood circulation, producing serious damage and often death. While the caissons are a few feet underwater, this danger is relatively minor, but at depths greater than thirty feet the pressure required to keep the water out of the working chamber is so high that upon rapid decompression the sand hog feels a sudden, acute pain in the stomach that bends him double. (Hence the nickname of the condition, the bends.) Paralysis follows.

Washington Roebling adopted caissons for the foundations of the bridge. The caisson on the Brooklyn side was 168 feet by 102 feet, made out of wood, with a roof twenty feet thick and walls decreasing from eight feet at the top to two inches at the cutting steel sheathing edge. It was to be founded forty-four feet under water. The New York side caisson was 172 feet by 102 feet with a roof twenty-two feet thick and was to be founded seventy-two feet under water. All kinds of accidents plagued the descent of the Brooklyn caisson. On December 2, 1870, a fire in it was caused by the flames of the lights or of a worker's candle that set fire to the oakum caulking of the caisson's walls. It had to be flooded and its sides and roof patched up. Another source of dangerous incidents was the miscalculated overpressure in the working chamber causing the air to bubble out of the caisson's bottom along the sides. Minor bubbles did not blow out much of the working chamber air, but major ones, like the one called the Great Blow-out, spouted rock, mud, and water to a height of 500 feet. Luckily the great Blow-out occurred on a Sunday at six in the morning, when nobody was in the chamber. The chamber had to be flooded again to compensate for the lost air pressure, which was also reduced when the air-locks were opened by mistake. Miraculously almost no damage occurred as the caisson dropped suddenly ten inches into the soil. But by the middle of 1875 the Brooklyn tower had reached its height of 316 feet, as its caisson had gone down at the rate of three feet per week thanks to the dynamite used to blast the big boulders found on the way down.

The New York caisson produced the first cases of the bends. One hundred-ten sand hogs were hit and three died during the first five months of 1872—a toll equal to the incidents in the caissons built by Eads for the Mississippi River Bridge in St. Louis. The doctor in charge of Eads's St. Louis Bridge had been struck by the bends himself, and was the first to realize their cause. He established a decompression rate of not more than six pounds of pressure per minute, a substantial improvement over the few minutes of decompression used by the sand hogs, but a far cry from the maximum of one pound of pressure reduction per minute in use now. With the rate adopted by Roebling, the danger of the bends for workers

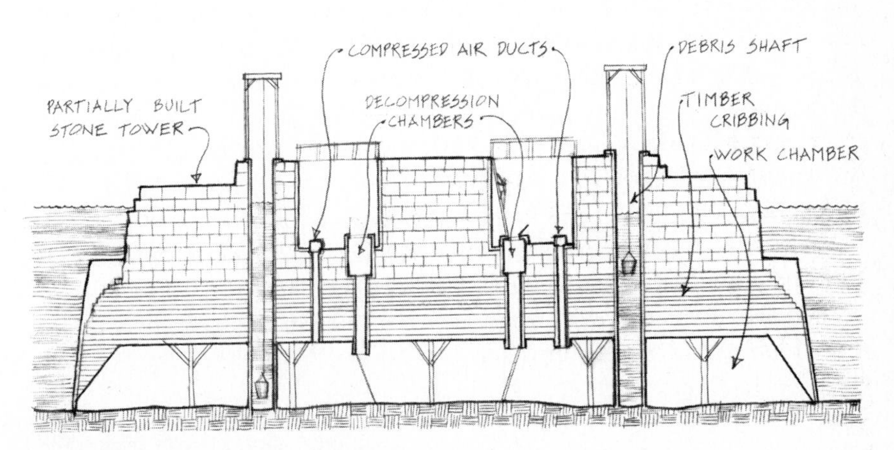

10.3 THE WORKING CHAMBER

at a depth of seventy-two feet was great indeed. Although Roebling reduced their shift to two hours and allowed only the strongest men to work at great depths, incidents of this type continued to occur, and finally the Chief Engineer, who was with his sand hogs so very often, was struck himself.

Paralyzed from the waist down, Washington found in his wife Emily the faithful interpreter of his ideas and the new leader of the construction crew. While he sat in his room overlooking the river from the Brooklyn side, Emily carried out his orders, handled the urgent daily problems, drew sketches, dealt with the workers, and became an expert on bridge construction, aware of every detail of this complex work. Forceful but shy by nature, she refused any and all honors in favor of her husband and at the end of construction retired to their house, where she died in 1903, leaving her husband and John, her child born in Europe during their "caisson-studying" trip. The sister of General Warren of Civil War fame, she was born and rests in Cold Spring, New York, one of the very first liberated women of the nineteenth century to dedicate some of her best energies to engineering.*

Finally by 1872 both caissons had reached bottom, the Brooklyn caisson on rock and the New York caisson on hard sand, seventy-nine feet under the river level. Caissons and wells were filled with concrete and the New York tower was completed in 1876, a year after the Brooklyn tower. They are 140 feet long and 59 feet wide at water level, and 136 feet long and 53 feet wide at their top. Each has two 119-feet-high and 34-feet-wide Gothic arches, where the roadway meets the towers. They are two most imposing monuments, contrasting the slenderness of the steel cables with their massive granite structure. Finally work on the cables could start. From laboring 79 feet under water the crew now proceeded to work at 266.5 feet above it, a height surpassed then only by the spire of Trinity Church.

Washington Roebling lost the bid for the cable wires to a competitor, the Haigh Wire Company of Brooklyn. Like his father before him and James Eads before his father, Washington did not trust the steel companies of the day and insisted on having each batch of wire tested in the factory before shipment to the site. With this guarantee he proceeded. He hoisted a wire rope from its anchorage in Brooklyn over the Brooklyn tower, ferried it across the river, hoisted it over the New

* The words on her tombstone say: "Emily Warren Roebling, Gifted, Noble, True. The wife of Washington A. Roebling." Not a word about the bridge.

York tower and anchored it on the New York side. Three more ropes were similarly strung and the spinning of the cables started from the aerial catwalks hanging from the ropes. Four heavy cast iron blocks, shaped like saddles, were set on top of the towers to guide the cables, and the crucible steel wires began crossing from Brooklyn to New York

10.4 E. F. Farrington Crossing the Bridge (from Harper's New Monthly Magazine)

and back. Two hundred seventy-eight wires constituted a bundle called a *strand* and nineteen strands constituted a cable, fifteen and three-quarters inches in diameter. Four cables were spun over four saddles, two close to the center line of the bridge and two along its outer edges.

The strength of the ropes supporting the catwalks had been spectacularly "tested" by the chief mechanic and assistant bridge engineer, E.F. Farrington, who crossed the river both ways on a bosun's chair hanging from a pulley over the first rope, to the acclaim of the populace, the first user of the Brooklyn Bridge (Fig. 10.4). But on June 14, 1878, a strand of wires broke loose from the New York anchorage, fell into the river, barely missed a ferryboat, killed one worker instantly, fatally injured another and hurt three more. A scandal developed and Roebling was accused of having used some weaker Bessemer steel from his own company. A thorough investigation proved instead that his rival's company had all along faked the strength tests on the wire by the simple expedient of testing a batch of good wire, loading it on a wagon for transportation to the site, unloading the wagon en route, and reloading it with inferior wire. This operation was repeated with successive wagon loads, using the same batch of good, tested wire each time. Roebling had 150 wires added to the cables to compensate for the strength deficiency of the delivered wire and the Haigh Wire Company was allowed to keep their contract.

Once the nineteen strands of each cable had been spun, they were wrapped in a tight spiral of coated steel wire, supplied by the Roebling Company. The wire spiral binds the strands to make them work together, gives the cables a circular shape, and prevents water from reaching the wire strands. Three hundred eighty suspenders of steel rope attached to the cables reach down to the level of the roadway deck, and stays of decreasing slope connect further points on the deck to the top of the towers. The stays are strong enough to support the weight of the deck in case of suspender or cable failure, a most unlikely event since, like all good engineers, John Roebling had been a most conservative man—the cables can carry a weight six times heavier than the weight of the roadway.

The work of seven years had only one purpose: to build a support for the roadway, over which the flow of trains, carriages, and pedestrians was to run. The roadway is supported by transverse steel trusses hanging from the suspenders and connected by longitudinal beams to constitute a steel grid. It was built by starting simultaneously at the two towers and moving towards the center of the bridge. The two vertical stiffening trusses along the edges of the bridge were erected on this grid. They are ten feet high and are connected at their top by transverse beams. The roadway thus has a rectangular pipe shape with sides built of trusses. This type of construction is very rigid, resists twisting and adds strength against dangerous twisting bridge oscillations due to wind. The original roadway had six lanes, two near the center line of the bridge for the railroad tracks and two double lanes on the outside for carriages (Fig. 10.5). In addition, a pedestrian path ran at a higher level along the centerline of the bridge. In 1948 D.B. Steinman, a world famous bridge engineer and an impassioned New Yorker, was entrusted with the strengthening of the bridge. The train lanes were converted to car traffic and additional light trusses were added over the side trusses (Fig. 10.5). These impede somewhat the view from the bridge, but do not impair its beauty. Together with the complex network of new approaches on the New York side, the almost 100-year-old bridge is fulfilling its mission as well or better today than it did in 1883.

Even the greatest works can be and are often misused by the unwise. The Brooklyn Bridge, dedicated to the union of two cities, fruit of the unswerving devotion of its engineers and workers, a catalyst to the erection of fifteen additional bridges connecting Manhattan to Brooklyn, the Bronx, and New Jersey, and Brooklyn to Staten Island, has become the favorite jumping platform for suicides. They fall 135

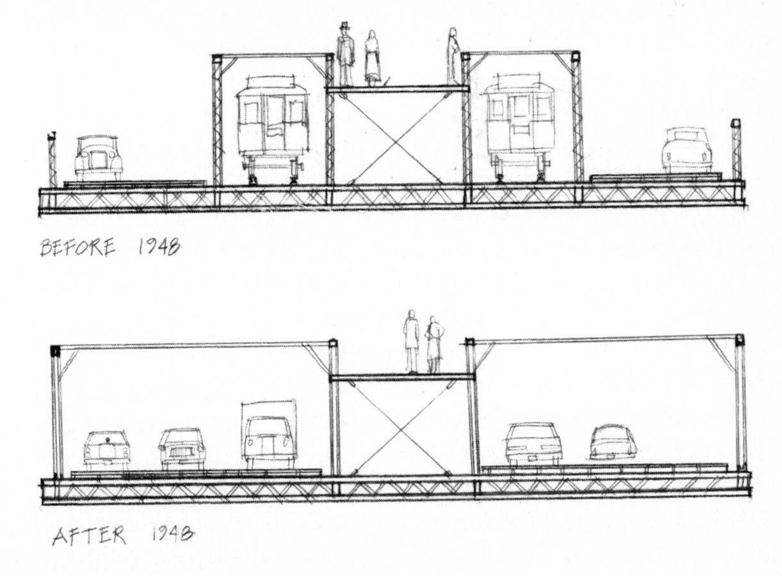

10.5 THE ROADWAY OF THE BRIDGE

feet, but even so some have been known to survive. Robert Odlum jumped first in 1885 and died. The legendary Steve Brodie, who boasted about his jump to the clients of his bar, probably never did jump. Stuntmen have also jumped with mixed results. But the largest number of people killed by the bridge in one day occurred only a week after its opening. Twelve were trampled to death in a panic perhaps created by someone who yelled that the bridge was falling. Their senseless deaths remind us of the tragic deaths of the more than twenty workers killed on the job and of the maiming of its creators. Emily was given only thirteen years to gaze upon the bridge that absorbed her labor for longer than that, but when Washington Roebling died, aged eighty-nine, in 1926, his health almost recovered, he had contemplated for over forty-three years one of the greatest achievements in the history of beneficial technology. He could well feel fulfilled: his father had conceived the bridge but he had built it.

Modern Suspension Bridges

Suspension bridges are the kings of the bridge world. No other method of construction can span greater distances. Their use of materials is totally logical. Steel cables of the highest tensile resistance developed to date in any man-made material are used in tension to support through the vertical or inclined suspenders of wire rope the weight of the roadway and the traffic load on it. The massive and inexpensive concrete of the anchorages, in which the cables are tied through steel anchor-bars to deeply embedded anchorplates, resists the pull of the cables by its weight. The steel towers, on top of whose saddles the cables are allowed to slide slightly to accomodate movements due to changes in temperature and loads on the roadway, are pushed down and compressed by the tension in the cables, as the center pole of a tent is compressed by the pull of the ropes. The longitudinal trusses of steel, like enormous beams, give rigidity to the roadway through their resistance to bending, so that the roadway does not move appreciably up and down under the moving loads. The harmonious collaboration of right materials, correctly shaped, makes a suspension bridge both a triumph of technology and a work of beauty. Gone are the days when Cass Gilbert, the architect of the Woolworth Building, insisted that the towers of the George Washington Bridge should be covered with stone. By a great consensus the structure of the steel towers was left uncovered to show with pride its obvious strength (Fig. 10.6).

10.6 THE TOWERS OF THE GEORGE WASHINGTON BRIDGE

Great strides have been achieved since the days of the Roeblings. The longest American bridge, the Verrazano Narrows Bridge at the entrance of New York Harbor, spans 4,260 feet between towers, a length over two-and-a-half times that spanned by the Brooklyn Bridge. But the Humber Bridge in Great Britain is 4,626 feet long and the Akashi-Kaikyo Bridge in Japan will span 5,800 feet in 1985. The Brooklyn Bridge used over 16,000 miles of cable wire. The Verrazano has 143,000 miles of it.

The improvements in suspension bridge design are not limited to an increase in dimensions and steel quantities. Many more subtle ideas have bettered their performance and reduced their cost. For example, both the George Washington Bridge and the Verrazano have two superimposed decks, which together with their side trusses give their decks a closed, rectangular pipe shape stronger against twisting than a single deck. On the other hand, the single deck of the Humber Bridge, which hangs from concrete towers, is a closed steel box with thin edges which splits the wind and prevents by its shape the tendency to excite twisting oscillations. These are also prevented by the suspenders that, instead of being vertical, are alternately inclined in opposite directions, thus partly acting as inclined stays (Fig. 10.7).

10.7 THE HUMBER BRIDGE

Elegant and light in appearance, economical in design, safe in erection, suspension bridges are spreading to all parts of the world, connecting cities, nations, and continents and keeping busy the few specialized crews of "wire spinners" from continent to continent. Their ultimate limits have been approached but not achieved yet and their future is secure. They are making more of our earth one world.

11 Form-Resistant Structures

Grids and Flat Slabs

Ever since the beginning of recorded history, and we may assume even earlier, people have gathered in large numbers for a variety of purposes be they religious, political, artistic, or competitive. The large roof, unsupported except at its boundary, arose to shelter these gatherings, evolving eventually into the huge assembly hall we know today.

As we shall see, no large roof can be built by means of natural or man-made compressive materials without giving the roof a curved shape, and this is why domes were used before any other type of cover to achieve large enclosed spaces. Even wood, a material that can span relatively short horizontal distances by beam action (see Chapter 5), has to be combined in conical, cylindrical or spherical shapes whenever large distances are to be spanned.

Only after the invention of inexpensive methods of steel manufacture and the recent development of reinforced concrete did large flat roofs become possible. They have obvious advantages over dome roofs: their erection is simpler, and they do not waste the upper part of the space defined by the dome which is often superfluous, unnecessarily heated or air-conditioned.

The simplest structural system for a flat rectangular roof consists of a series of parallel beams supporting some kind of roofing material. But if all four sides of the rectangle to be covered can be used to support the roof beams, it becomes more practical to set the beams in two directions, at right angles to each other, thus creating a *grid.* This *two-way system* pays only if the two dimensions of the rectangle are more or less equal. Loads tend to move to their support through the shortest possible path and if one dimension of the roof is much larger than the other, most of the load will be carried by the shorter beams, even if the beams are set in a grid pattern.

A grid is a "democratic" structural system: if a load acts on one of its beams, the beam deflects, but in so doing carries down with it all the beams of the grid around it, thus involving the carrying capacity of a number of adjoining beams. It is interesting to realize that the spreading of the load occurs in two ways: the beams parallel to the loaded beam bend together with it, but the beams at right angles to it are also compelled to twist in order to follow the deflection of the loaded beam (Fig. 11.1). We thus find that in a rectangular grid loads are carried to

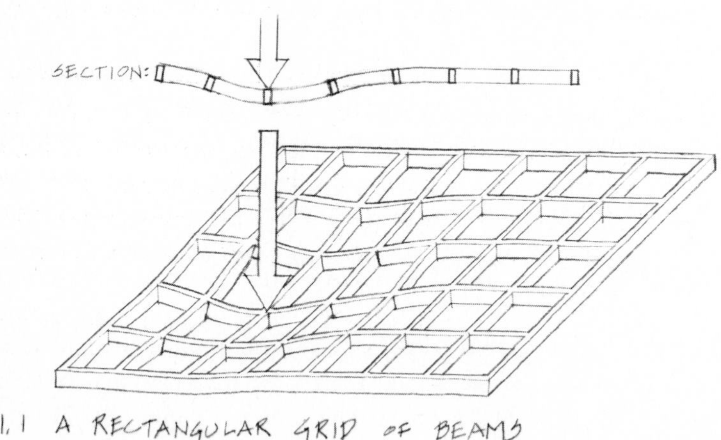

11.1 A RECTANGULAR GRID OF BEAMS

the supports not only by beam action (bending and shear) in two directions but by an additional twisting mechanism which makes the entire system stiffer. To obtain this twisting interaction the beams of the two perpendicular systems must be rigidly connected at their intersection, something which is inherent in the monolithic nature of reinforced concrete grids and in the bolted or welded connections of steel grids. Even primitive people know how to obtain such twisting action by interweaving the beams of their roofs so that any displacement of one beam entails the bending and twisting displacement of all the others (Fig. 11.2).

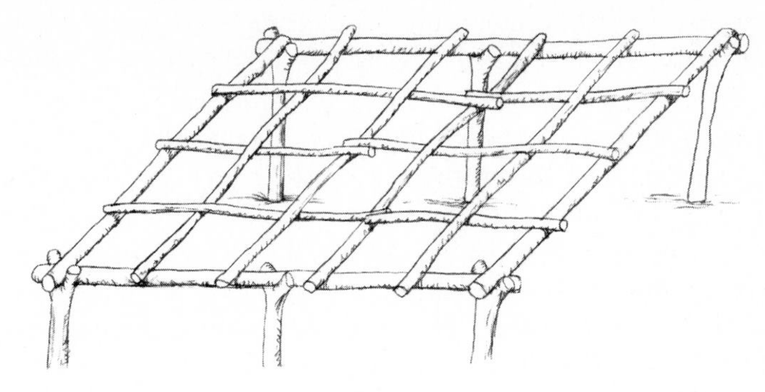

11.2 A WOVEN GRID OF BEAMS

Though rectangular grids are the most commonly used, *skew grids* (Fig. 11.3) have, beside aesthetic qualities, the structural and economic advantage of using equal length beams even when the dimensions of the grid are substantially different, thus distributing more evenly the carrying action between all the beams.

We have seen in Chapter 9 how grids of trusses rather than beams become necessary when spans are hundreds of feet long, and how space frames constitute some of the largest horizontal roofs erected so far, covering four or more acres without intermediate supports. We must now go one step back to discover how an extension of the grid concept has become the principle on which most of the floors and roofs of modern buildings are built.

Let us imagine that the beams of a rectangular grid are set nearer and nearer to each other and glued along their adjacent vertical sides

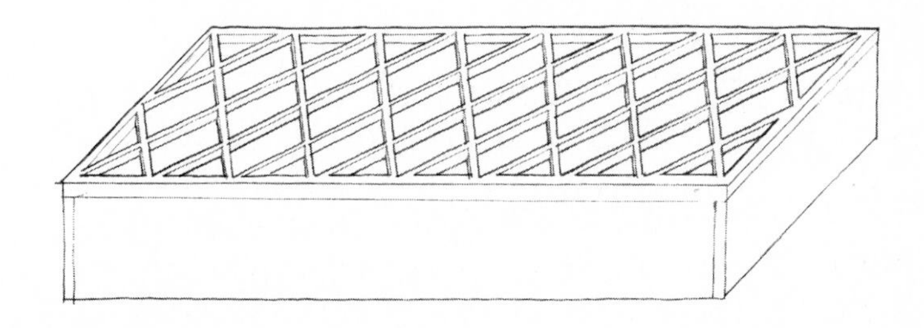

11.3 A SKEW GRID

until they constitute a continuous surface. Such a continuous surface, called a *plate* or *slab*, presents all the advantages of a grid in addition to the ease with which it can be poured on a simple horizontal scaffold when made out of concrete. Reinforced concrete horizontal slabs are the most commonly used floor and roof surfaces in buildings with both steel and concrete frames all over the world. Their smooth underside permits a number of things to hang—pipes and ducts, for instance—without having to go around beams. The setting of the slab reinforcement on flat wooden scaffolds makes the placing of the steel bars simple and economical. In European countries concrete slabs are sometimes made lighter by incorporating hollow tiles (Fig. 11.4). Through the strength of their burnt clay these tiles participate in the slab structural action, which is the same in all slabs whatever their material.

Actually slabs, besides carrying loads by bending and twisting like grids of beams, have an additional capacity which makes them even stiffer and stronger than grids. This easily undersoood capacity derives from the continuity of their surface. If we press on a curved sheet of material attempting to flatten it, depending on its shape, the sheet will flatten by itself or have to be stretched or sliced before it can be made flat. For example, a sheet of paper bent into a half-cylinder and then re-

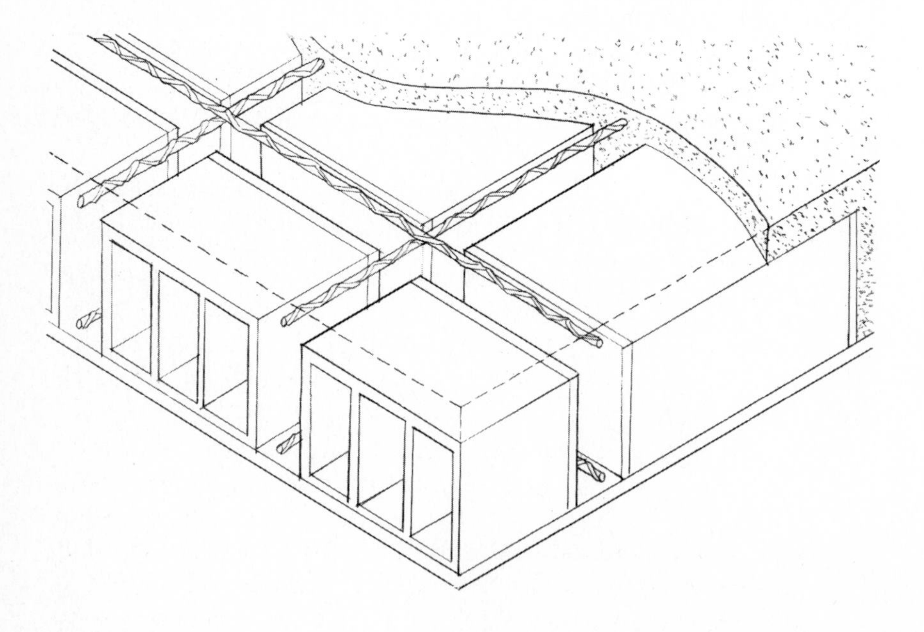

11.4 TILE-CONCRETE FLOOR SLAB

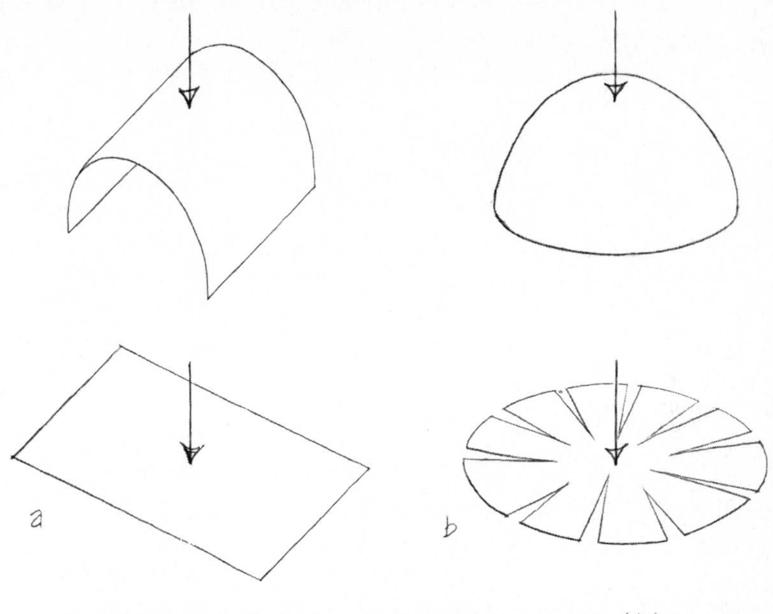

11.5 DEVELOPABLE (a) AND NON-DEVELOPABLE (b) SURFACES

leased flattens by itself (Fig. 11.5a). It is said to be a *developable surface* (from the idea "to unfold" contained in the verb "to develop"). But if we cut a rubber ball in half, producing a small spherical dome, the dome will not flatten by itself if we lay it on a flat surface. Neither will it become flat if we push on it. It only flattens if we cut a large number of radial cuts in it or if, assuming it is very thin, it can be *stretched* into a flat surface (Fig. 11.5b). The dome (and actually all other surfaces except the cylinder) are *non-developable, unflattening surfaces.* Because they are so hard to flatten, they are also much stiffer than developable surfaces. (It will be more obvious why non-developable surfaces are better suited to build large roofs once we learn how such roofs sustain loads.)

Returning now to the behavior of a flat slab, we notice that under load it becomes "dished"—it acquires the shape of a curved surface, with an upward curvature (Fig. 11.6). If it is supported only on two opposite parallel sides, it becomes a slightly curved upside-down cylinder, but if it is supported on four sides, or in any other manner, it acquires a non-developable shape. Just as the half-ball had to be stretched to be changed from a dome to a flat surface, the plate has to be stretched to

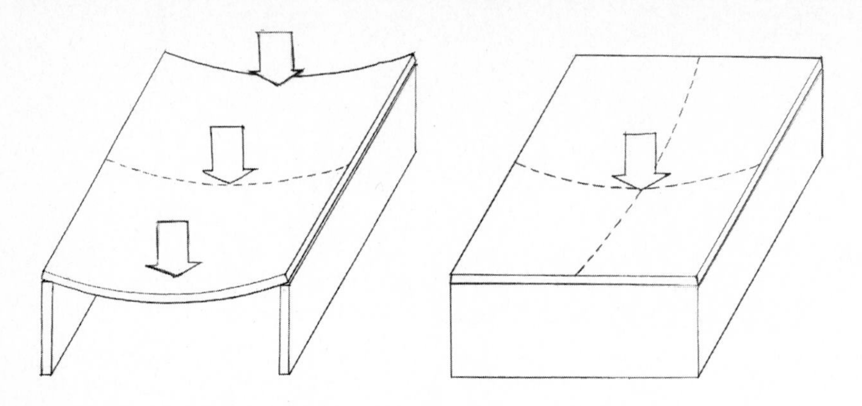

11.6 FLAT SLABS DISHED BY LOADS

change it from a flat to a dished surface. Hence the loads on it, besides bending it and twisting it, must stretch it, and this unavoidable stretching makes the slab even stiffer. Therefore we should not be amazed to learn that plates or slabs can be made thinner than beams. While a beam spanning twenty feet must have a depth of about one-and-one-half feet, whether it is made of steel, concrete or wood, a concrete slab covering a room twenty-feet square can be made one foot deep or less.

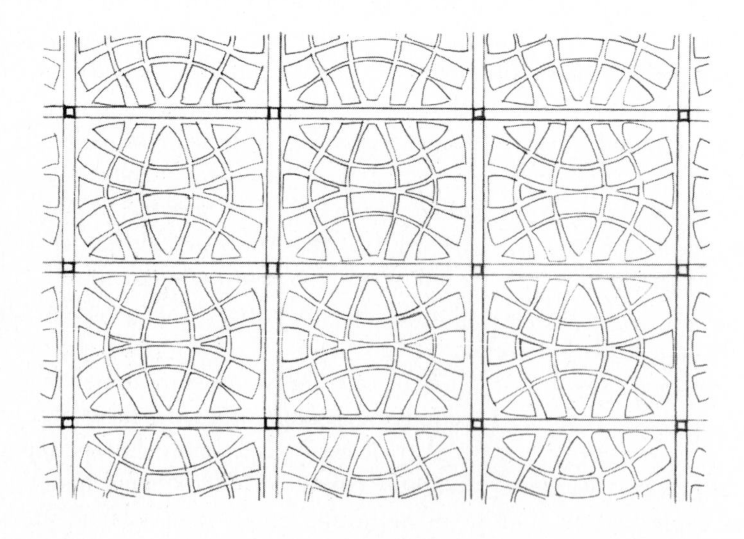

11.7 NERVI'S SLAB WITH CURVED RIBS

When slabs have to span more than fifteen or twenty feet, it becomes economical to stiffen them on their underside with ribs, which can be oriented in a variety of ways. Nervi made use of Ferrocemento, a material he perfected, to build forms in which to pour slabs stiffened by *curved* ribs, which are oriented in the most logical directions to transfer the loads from the slab to the columns. These curved ribs, moreover, give great beauty to the underside of the slabs (Fig. 11.7). Ferrocemento is a material consisting of a number of layers of welded mesh set at random, one on top of the other, and permeated with a concrete mortar, a mixture of sand, cement, and water (Fig. 11.8). Flat or curved elements of Ferrocemento can be built only one or two inches thick, with excep-

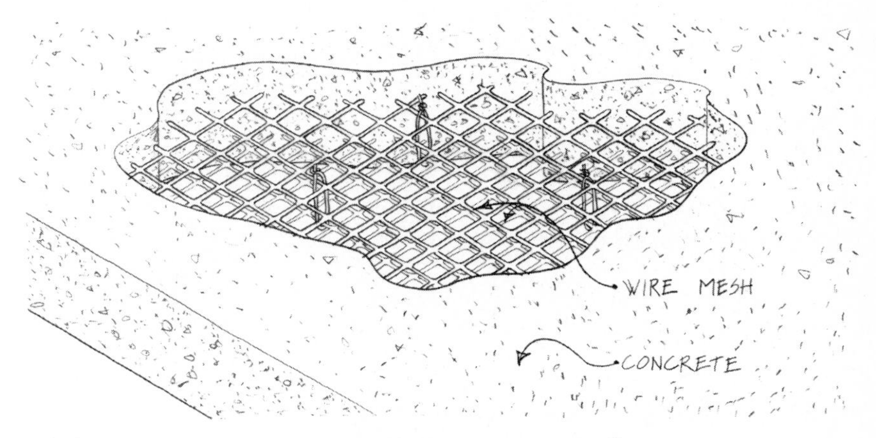

11.8 FERROCEMENTO MESH REINFORCEMENT

tional tensile and compressive strength due to the spreading of the tensile steel mesh through the high-strength compressive mortar. First used only as a material to build complex molds in which to pour reinforced concrete elements, it later was transformed by Nervi into a structural material itself. Some of the masterpieces of Nervi owe their extraordinary beauty and efficiency to the use of Ferrocemento.

Genius often consists of an ability to take the next step, and Nervi took it by realizing that Ferrocemento would be an ideal material for building boats. His lovely ketch *Nennele* (Fig. 11.9) was the first, but a large number of sailing boats have been built, mostly in Australia and

11.9 NERVI'S KETCH NENNELE (OF FERROCEMENTO)

the United States, with Ferrocemento hulls. They are easy to manufacture and even easier to repair in case of an accident.

Strength through Form

The stiffness of flat slabs, like that of beams, derives from their thickness: if too thin they become too flexible to be functional. It is one of the marvels of structural behavior that stiffness and strength of sheet-like elements can be obtained not only by increasing their thickness and hence the amount of required material, but by giving them curved shapes. Some of the largest, most exciting roofs owe their resistance exclusively to their shape. This is why they are called *form-resistant structures.*

If one holds a thin sheet of paper by one of its short sides, the sheet is incapable of supporting even its own weight—the paper droops down (Fig. 11.10a). But if we give the side held a slight curvature up, the same sheet of paper becomes stiffer and capable of supporting as a cantilever beam not only its weight but also the small additional weight of a pencil or pen (Fig. 11.10b). We have not strengthened the paper sheet by adding material to it; we have only curved it up. This principle of strength through curvature can be applied to thin sheets of reinforced

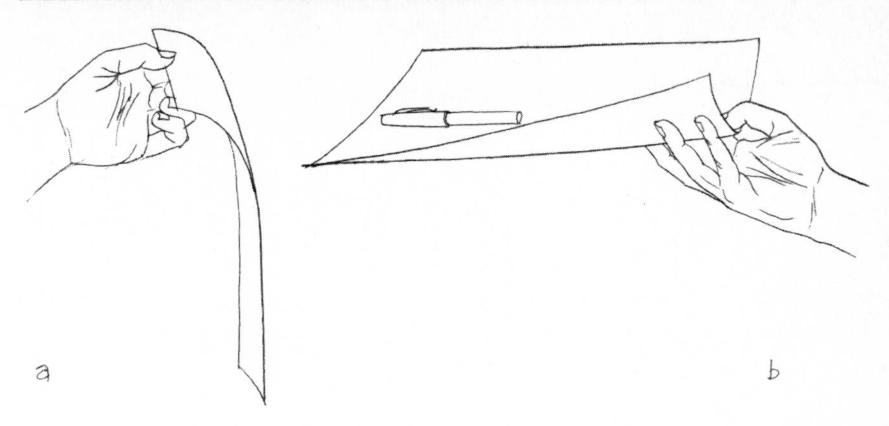

11.10 PAPER SHEET STIFFENED BY CURVATURE

concrete and has been efficiently used to build stadium roofs that may cantilever out thirty or more feet with a thickness of only a few inches (Fig. 11.11). The shape of such roofs can be shown to be non-developable and hence quite rigid, but even developable surfaces, like cylinders, show enough strength (when correctly supported) to allow their use as structural elements. To demonstrate this property, try to span the distance between two books by means of a flat sheet of paper acting as a plate. The paper will sag, fold, and slide between the book supports. If instead the sheet of paper is curved up and prevented from spreading by the book covers, it will span the distance as an arch (Fig. 11.12). Again the curvature has given the thin paper its newly acquired stiffness and strength.

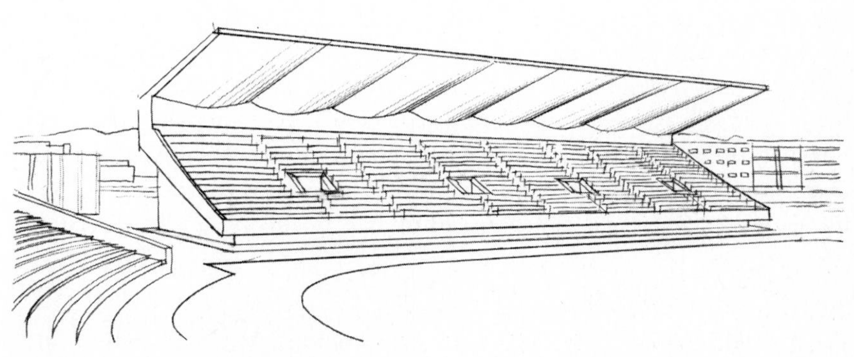

11.11 STADIUM STANDS ROOF

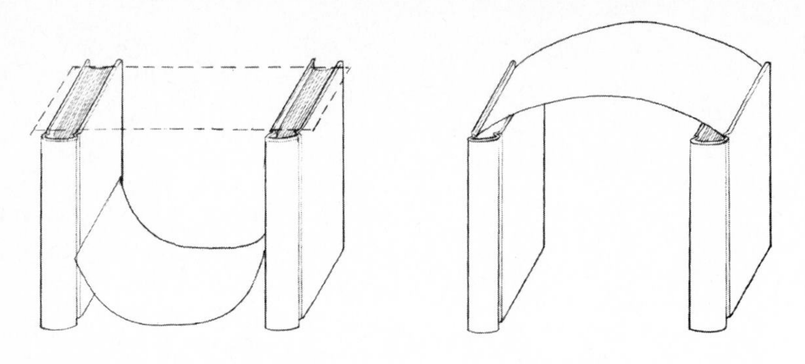

11.12 PAPER SHEET STIFFENED BY CYLINDRICAL SHAPE

Nature knows well the principle of strength through curvature and uses it whenever possible to protect life with a minimum of material. The egg is a strong home for the developing chick, even though its shell weighs only a fraction of an ounce. The seashell protects the mollusk from its voracious enemy and can, in addition, sustain the pressure of deep water thanks to its curved surfaces. The same protection is given snails and turtles, tortoises and armadillos, from whom our medieval knights may have copied their curved and relatively light armor.

Curved Surfaces

We owe to the greatest of all mathematicians, Karl F. Gauss (1777–1855), the discovery that all the infinitely varied curved surfaces we can ever find in nature or imagine belong to only three categories, which are domelike, cylinderlike, or saddlelike.*

How do the three categories differ? Consider the dome. Imagine cutting it in half vertically with a knife. The shape of the cut is curved downwards, and if you cut the dome in half in any direction, as you do when you cut a number of ice-cream-cake wedge slices, the shape of all the cuts is still curved downward (Fig. 11.13). A domelike surface has downward curvatures in all its radial directions. By the way, if instead of cutting a dome we were to cut in half a soup bowl, we would find the shape of all the cuts to be curved up, whatever their radial direction.

* Gauss was so great a man that he noted in a small book a number of discoveries "not worth publishing." When this booklet was found fifty years after his death, some of his "negligible" discoveries had been rediscovered and had made famous a number of his successors!

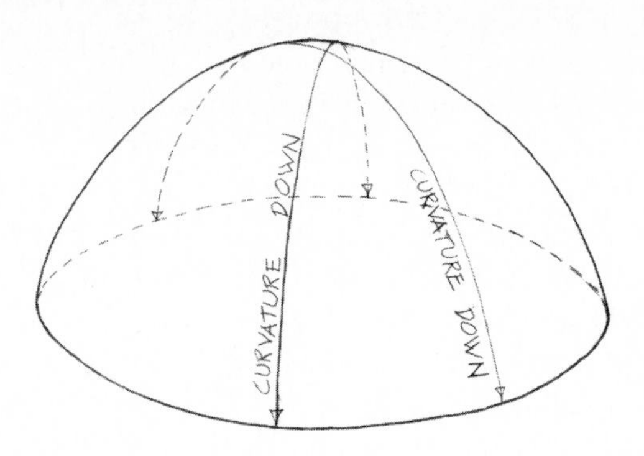

11.13 VERTICAL CUTS IN DOME

Domes and hanging roofs, each with curvatures always in the same direction (either down or up), constitute the first of Gauss's categories. They are non-developable surfaces and have been used for centuries to cover large surfaces. We will discuss their structural behavior in Chapters 13 and 15.

Let us jump to the third of Gauss's categories, the saddlelike surfaces. In a horse saddle the curve across the horse, defined by the rider's legs, is curved downward, but the curvature along the horse's spine, which prevents the rider from sliding forward or backward, is upward (Fig. 11.14). Saddle surfaces are non-developable and are used as roofs because of their stiffness. The Spanish architect Felix Candela built as a saddle surface what is perhaps the thinnest concrete roof in the world. Covering the Cosmic Rays Laboratory in Mexico City, it is only half-an-inch thick.

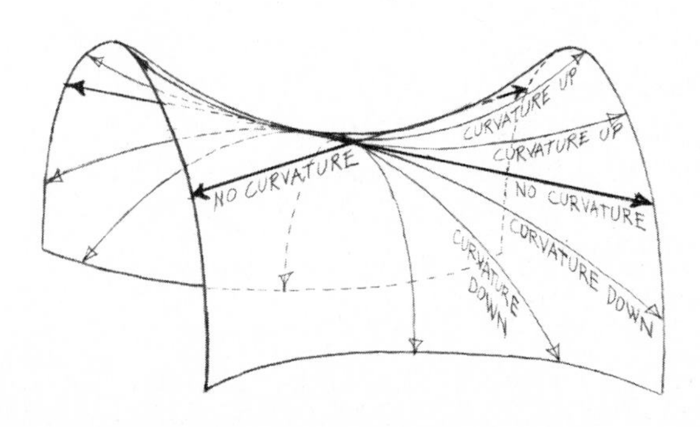

11.14 VERTICAL CUTS IN SADDLE

Saddle surfaces have another property not immediately noticeable. As one rotates the saddle cuts from the direction across the horse to that along the horse, the curvature changes from down to up and, if one keeps going, it changes again from up to down. Therefore there are two directions along which the cuts are neither up nor down. They are not curved; they are straight lines (Fig. 11.14). To prove this one has only to take a yardstick and place it across a saddle at its lowest point: the saddle is curved down, below the yardstick. If one then rotates the yardstick, keeping it horizontal, one finds that there is a direction along which *the yardstick lies entirely on the surface of the saddle:* in this direction the saddle has no curvature. Of course, if one rotates the yardstick in the opposite direction one locates the other no-curvature section of the saddle, which is symmetrical to the first with respect to the horse's axis. All saddle surfaces have two directions of no curvature. Cut along these directions, their boundaries are straight lines. This property makes the saddle shape an almost ideal surface with which to build roofs.

We can now go back to Gauss's second category, the cylinders. Imagine a pipe lying on the floor. If you cut vertically its top half—the half, say, with the shape of a tunnel—in any direction, you will notice that *all* of these cuts have a curvature down, *except one*: the cut along the pipe's axis is a straight line (Fig. 11.15). The cylinder has no curvature in the direction of its axis. One may consider, then, the cylinder as a dividing line between the dome and the saddle. The saddles have two directions without curvatures, but as these two directions draw nearer and nearer, saddles become cylinders, with only one direction of no curvature. If this direction is now given a down curvature, the cylinder becomes a dome. If instead of considering the upper part of a cylinder,

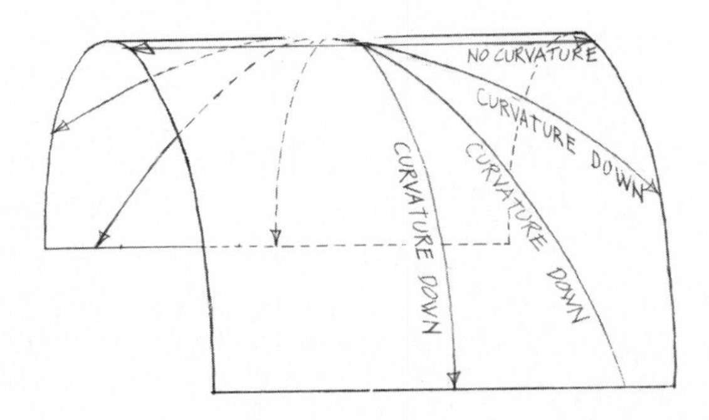

11.15 VERTICAL CUTS IN TOP HALF OF CYLINDER

we consider its lower half—the half, say with the shape of a gutter—we find that the vertical sections of the gutter have curvatures up in all directions except one: the direction of the axis of the gutter. Hence, gutters and tunnels belong to the same category of surfaces having only one direction of no curvature.

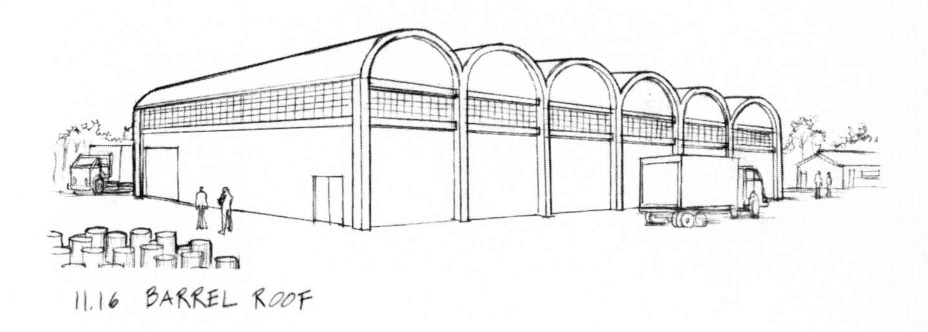

11.16 BARREL ROOF

Barrel Roofs and Folded Plates

We have seen that cylinders are developable surfaces and, as such, are less stiff than either domes or saddles. Even so, they can be used as roofs. Actually *barrel roofs* of reinforced concrete in the shape of half-cylinders with curvatures down are commonly and inexpensively used in industrial buildings (Fig. 11.16), since they can be poured on the same cylindrical formwork, which can be moved from one location to another and reused to pour a large number of barrels on the same form.

The mode of support of a barrel influences its load-carrying action. If a barrel is supported all along its two longitudinal edges (Fig. 11.17a), it acts as a series of arches built one next to the other and develops outpushing thrusts, which must be absorbed by buttresses or tie-rods as in any arch. But if it is supported on its curved ends (Fig. 11.17b), it behaves like a beam, developing compression above the neutral axis and tension below (see Chapter 5), and it does *not* develop thrusts. One should not be fooled by the geometrical shape of a structure in deciding its load-carrying mechanism. Barrels should be supported on end-walls or stiff arches so as to avoid unnecessary and costly buttresses or interfering tie-rods.

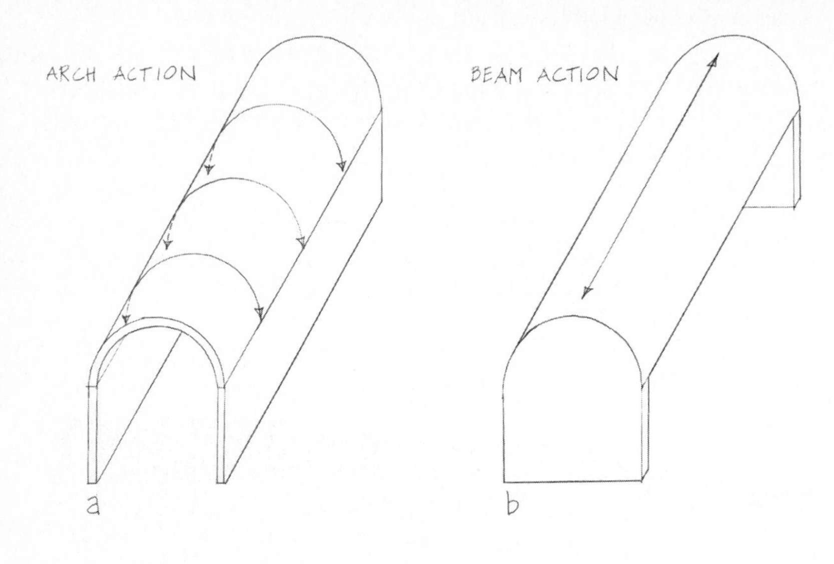

11.17 BARREL ROOF SUPPORTS

The *folded plate* roof is analogous to a series of barrels. It consists of long, narrow inclined concrete slabs, but presents a sudden *fold* or change in slope at regular intervals (Fig. 11.18). Its cross-section is a zigzag line with "valleys" and "ridges." The construction of a folded plate roof requires practically no formwork, since the flat slabs can be poured on the ground and jointed at the valleys and ridges of the roof by connecting the transverse reinforcing bars of the slabs and using a good cement grout or mortar to make the slabs into a monolithic roof.

11.18 FOLDED-PLATE ROOF

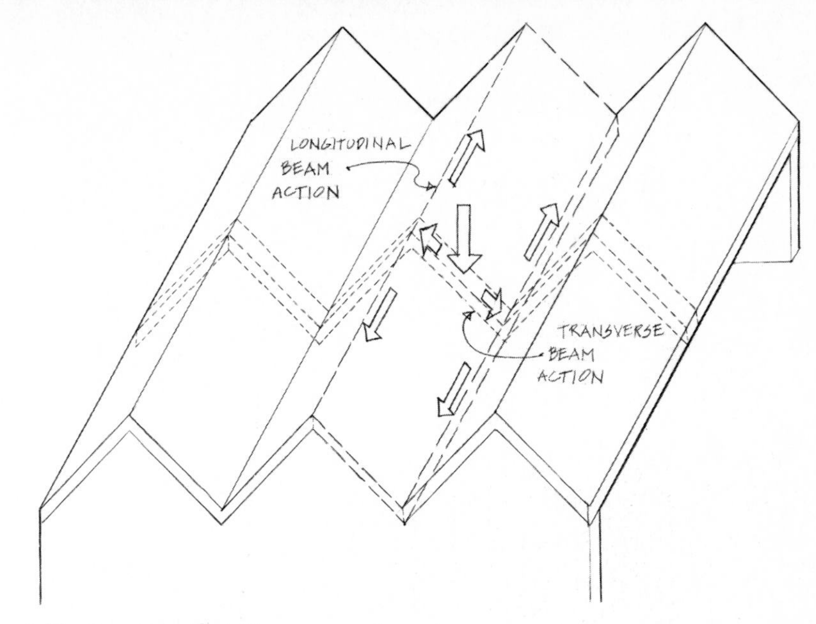

11.19 FOLDED-PLATE LOAD PATHS

Folded plates carry loads to the supports along a twofold path. Because of the stiffness achieved by the folds, any load acting on a slab travels first up the nearest ridge or down the nearest valley, and then is carried to the end supports longitudinally by the slabs acting as beams (Fig. 11.19). Folded plates must be supported at their ends. Since they consist of flat surfaces and folds, they act like an accordion that can be pushed in or pulled out with little effort, and do not develop out-pushing thrusts.

1.20 FOLDED-PLATE PAPER MODEL

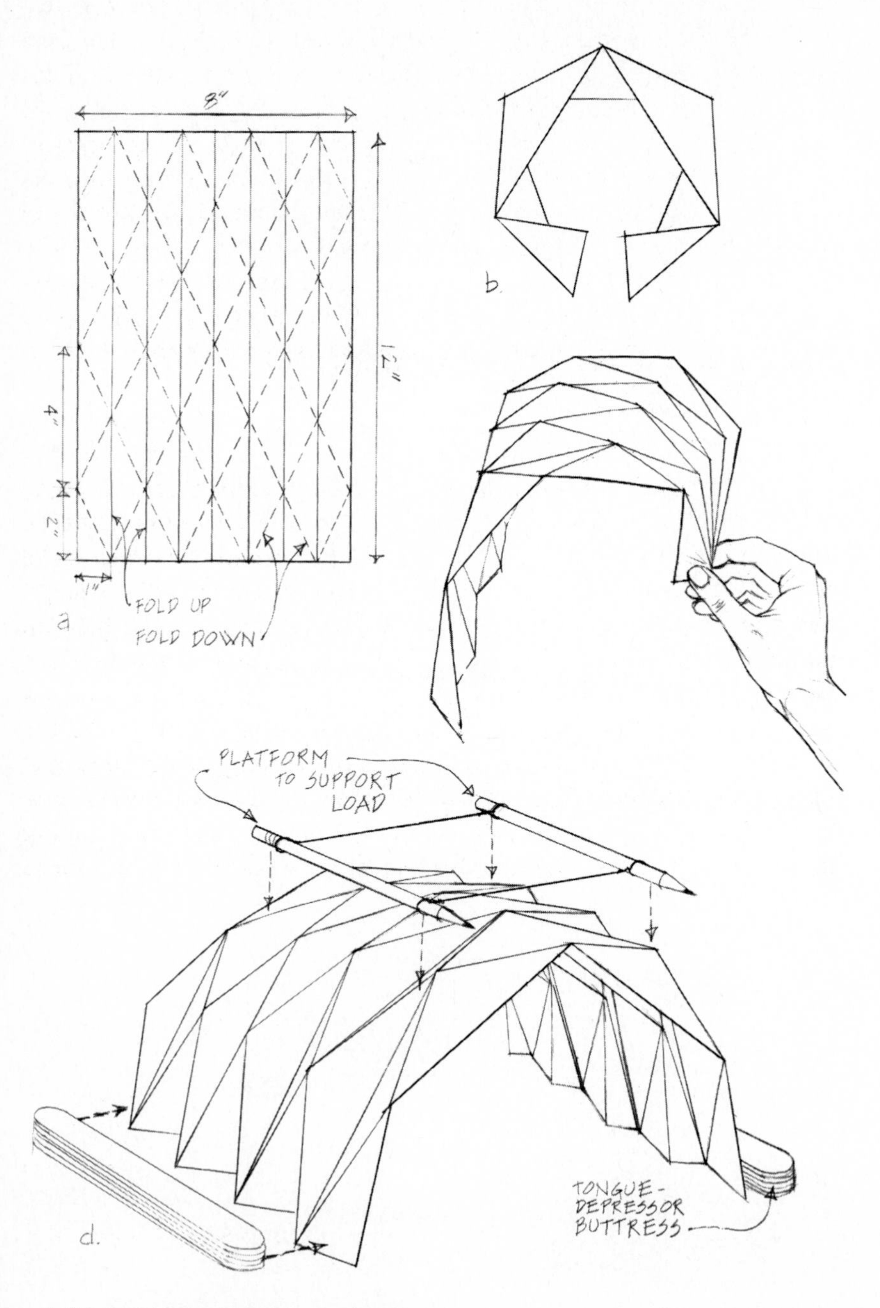

11.21 CREASED PAPER BARREL

It is both easy and instructive to fold a sheet of thin paper up and down, shaping it into a folded plate, and to support it between two books, possibly laying a flat sheet of paper over it (Fig. 11.20). The load capacity obtained by such a flimsy piece of material through its folds is amazing: a sheet of paper weighing less than one-tenth of an ounce may carry a load of books two or three hundred times it own weight! Any reader inclined to experiment further with folded paper can take advantage of both folding and arch action by creasing a sheet of paper into a folded barrel, according to the instructions of Figure 11.21. The creased paper barrel requires buttresses to absorb its outward-acting thrusts, but its load-carrying capacity is even greater than that of a folded-plate roof and may easily reach 400 times its own weight.

Saddle Roofs

Saddle surfaces, supported along their longitudinal curved edges, have a particularly elegant shape which blurs the distinction between structure and functional skin (Fig. 11.22). But saddle surfaces make some of the loveliest roofs when cut and supported along those straight lines which we have seen necessarily exist on any surface with both up and down curvatures (see Fig. 11.14). To visualize how a curved surface can be obtained by means of straight lines, connect by inclined straight-line segments the points of two equal circles set one above the other (Fig 11.23). The segments generate a curved surface called a *rotational hyperboloid*, used to build the enormous cooling towers of chemical plants. One of the most commonly used roof surfaces is obtained in a similar manner. Imagine a rectangle of solid struts, in which one of the corners

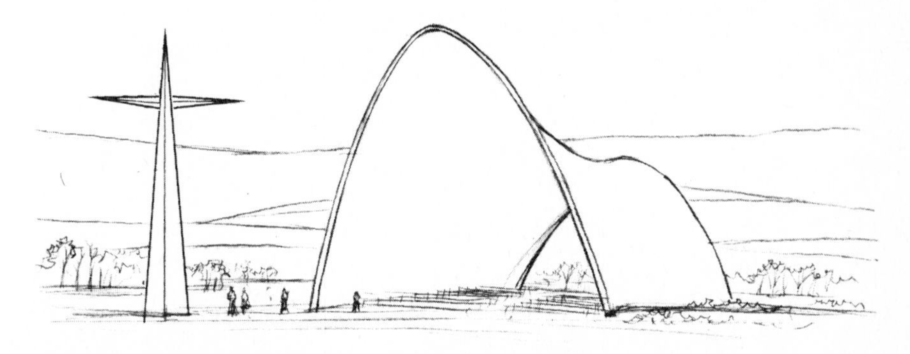

11.22 CANDELA'S SADDLE-SHAPED CHAPEL IN MEXICO

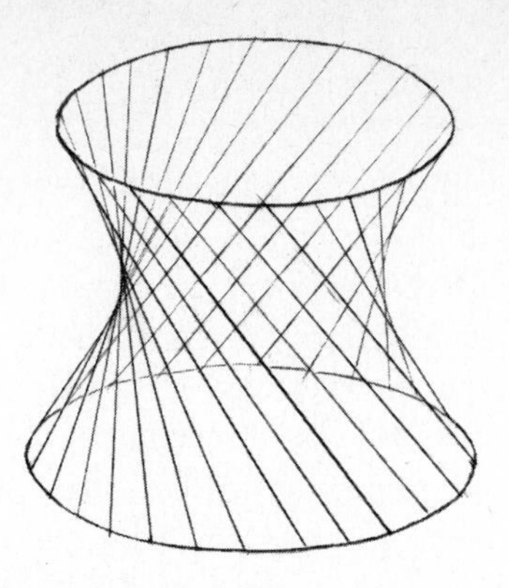

11.23 ROTATIONAL HYPERBOLOID

is lifted from the plane of the other three, thus creating a frame with two horizontal and two inclined sides (Fig. 11.24a). If the corresponding points of two opposite sides of this frame (one horizontal and one inclined) are connected by straight lines, for example by pulled threads, and the same is done with the other two opposite sides, the threads will describe a curved surface, although, being tensed, they are themselves straight (Fig. 11.24b). This surface has a curvature up along the line connecting the lifted corner to its diagonally opposite corner, and a curvature down in the direction of the line connecting the other two corners (Fig. 11.25). It is, therefore, a saddle surface. It carries the

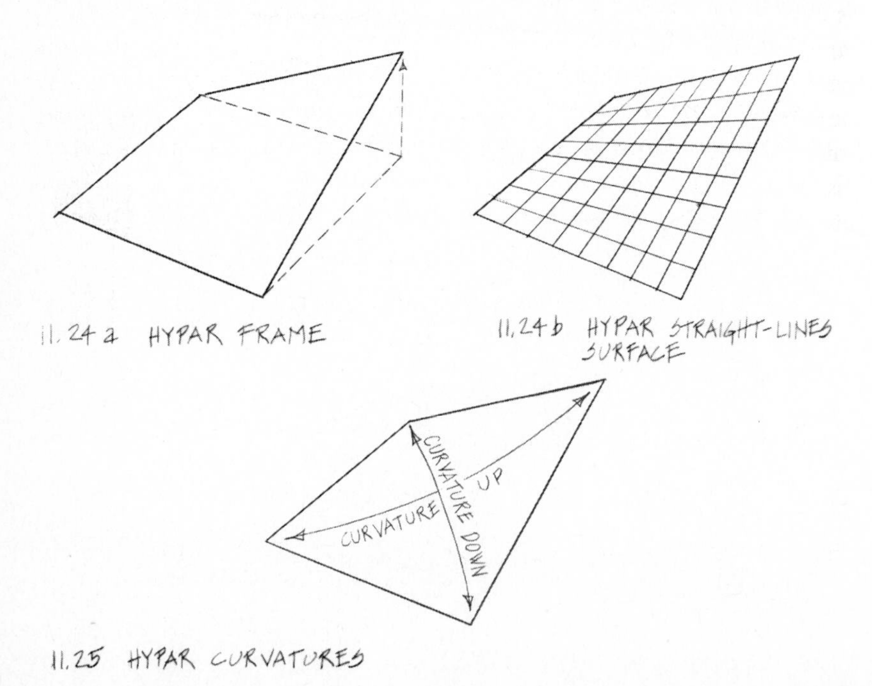

11.24a HYPAR FRAME

11.24b HYPAR STRAIGHT-LINES SURFACE

11.25 HYPAR CURVATURES

high-sounding name of *hyperbolic paraboloid,* wisely shortened to *hypar* by our British colleagues.

One of the simplest hypar roofs is obtained by tilting the saddle and supporting it on two opposite corners. Whether the support points are on the ground or on columns, the roof looks like a butterfly ready to take off (Fig. 11.26). Its structural behavior is dictated by its curvatures. Compressive arch action takes place along the sections curved downward and tensile cable action along the sections curved upward. The two support points must be buttressed to resist the thrusts of the arch action, while the tensile cable action at right angles to it must be absorbed by reinforcing bars, if the hypar is made out of concrete. Such is the stiffness of a hypar that its thickness need be only a few inches of concrete for spans of thirty or forty feet. The hypar has other wonderful structural properties. For example, one could fear that such a thin structure, acting in compression along its arched direction, would easily buckle, a fear quite justified were it not that the cable action at right angles to the arches pulls them up and prevents them from buckling! Finally, to make the structural engineer even more enamoured of these surfaces, under a uniform load, like its dead load or a snow load, they develop the same tension and compression everywhere. Therefore, its material, be it concrete or wood, can be used to its greatest allowable capacity all over the roof. The reader, who might not have seen too many of these magnificent roofs, may ask "Why do we not see many more of them?" The answer to this question is that there is no silver lining without a cloud and the cloud that hangs over the hypars is the cost of their formwork, as it is for all curved surfaces. More will be said later about this problem.

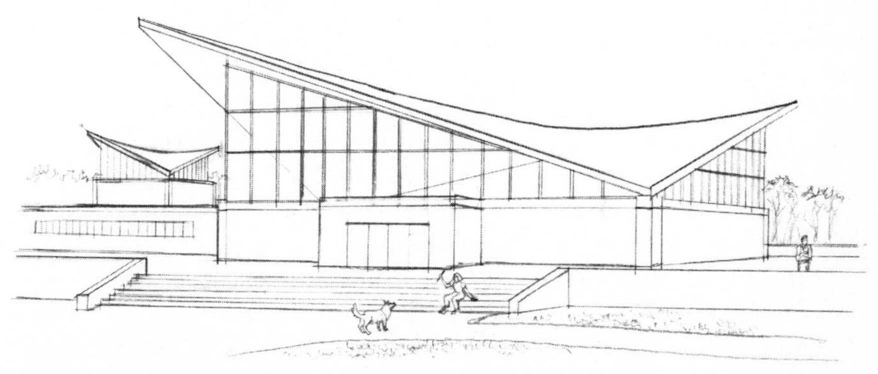

11.26 HYPAR BUTTERFLY ROOF

Complex Roofs

The barrel and the rectangular hypar elements are the building blocks for some of the most exciting curved roofs conceived by man. Combinations of these structurally efficient components are limited only by the imagination of the architect, guided by good structural sense. It is indeed regrettable that, with a few notable exceptions, modern architecture has not used curved surfaces as glorious as some of the past and at the same time as daring as present day technology can make them. This lack of achievement is due to at least three causes. On one hand, curved surfaces are believed to be more complex to design than the flat rectangular shapes we are so used to. Usually quite the opposite is true. On the other hand, there is a gap between recent curved-structures theory and the prescriptions of the codes. A domed roof proposed for a bank in California—meant to cover a rectangular area ninety feet by sixty feet and to be only a few inches thick—was vetoed by the local building department engineer because thin curved roofs were not mentioned in the code and, hence, "did not exist." The engineer would only allow the roof to be built if two concrete arches were erected between its diagonally opposite corners "to support it." Little did he know that the thin concrete roof was so stiff that *it* would support the two heavy arches rather than be supported *by them!* Finally, one must honestly add that in the United States the ratio of labor to material costs often makes *thin shells* (as these curved roofs are usually labelled) uncompetitive with other types of construction. The situation is reversed in Europe and other parts of the world.

One of the most commonly encountered combinations of cylindrical surfaces is the *groined vault* of the Gothic cathedrals (Fig. 11.27). This consists of the intersection of two cylindrical vaults at right angles to each other, supported on four boundary arches and intersecting along curved diagonal folds called *groins,* which end at the four corner columns supporting the vault. The groins have often been emphasized visually and, possibly, structurally by means of ribs but, though these ribs may be aesthetically important, they are not needed to sustain the vaults. By their curvature and folds they are self-supporting.

Among the great variety of combinations of rectangular hypar elements, two have become quite common because of their usefulness, beauty, and economy: the *hypar roof* and the *hypar umbrella.* To put together a hypar roof (Fig. 11.28a), consider building four hypar rec-

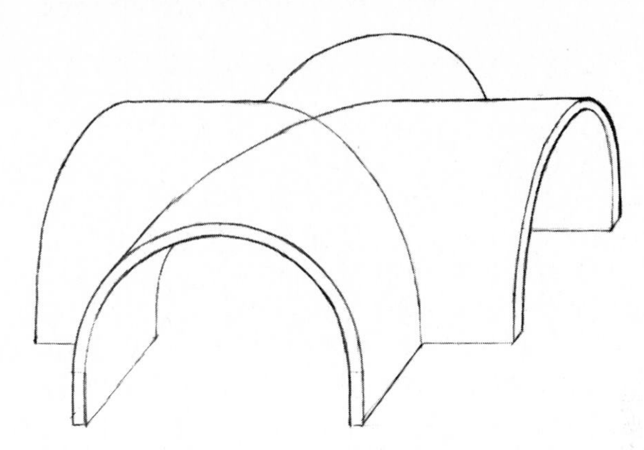

11.27 GROINED VAULTS

tangular elements, starting with four rectangles but *lowering* (rather than lifting as was done before) one corner in each of them. The hypar roof is obtained by joining together the horizontal sides of each rectangular element so that all eight meet at the center of the area to be covered, while the lowered corners are supported on four columns or on the ground at the corners of the area. The straight inclined sides of the roof act as the compressed struts of a truss, and they must be prevented from spreading outward by means of tie-rods connecting its corners, all around the covered area. The largest roof of this kind has been erected in Denver, Colorado, and measures 112 feet by 132 feet, with a three-inch thickness. It rests directly on the ground at the corners and covers a large department store.

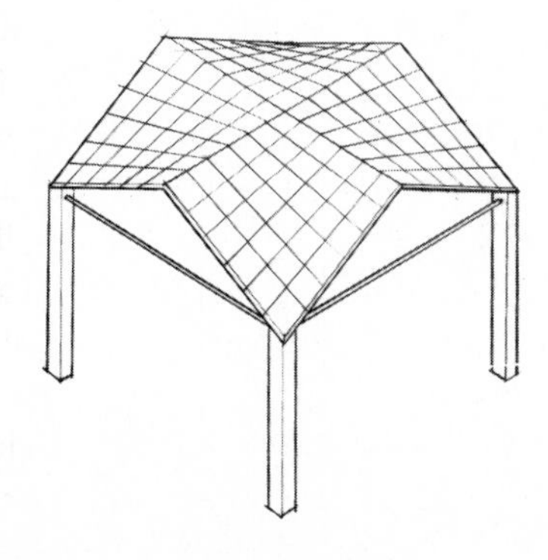

11.28a HYPAR ROOF

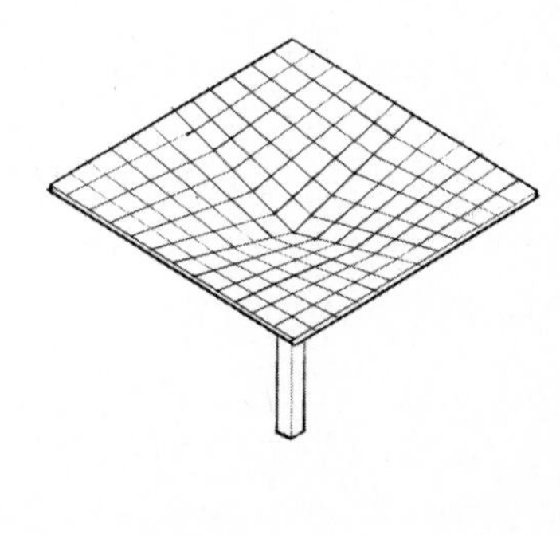

11.28b HYPAR UMBRELLA

11.29 HYPAR UMBRELLAS OF NEWARK AIRPORT

The hypar umbrella (Fig. 11.28b), one of the most elegant roof structures ever devised, is produced by using four rectangular hypar elements, each with a corner lowered with respect to the other three, and put together by joining the two inclined sides of each rectangular element so that all eight meet at the center of the area to be covered. The horizontal sides constitute the rectangular edge of the roof and hide the tie-rods. The shape of this hypar roof, which starts at a central point and opens up, resembles that of a rectangular umbrella and gives a visual impression of floating upward. Hypar umbrellas up to ninety feet square have been used, for example, in the terminal building of Newark Airport (Fig. 11.29).

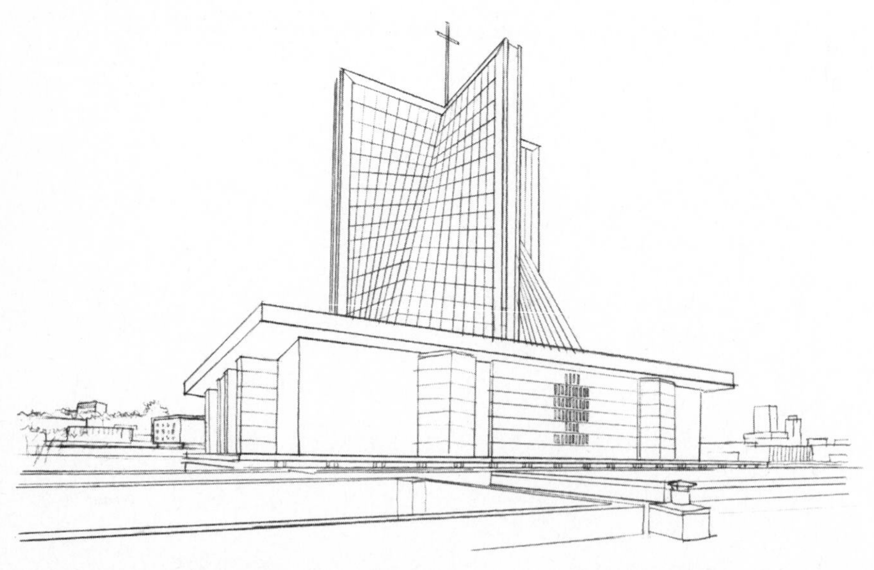

11.30 SAN FRANCISCO CATHEDRAL

Such is the variety of shapes which can be composed by means of hypars that the Spanish architect Felix Candela has become famous all over the world by designing and building, mostly in Mexico, roofs that use only this surface as basic element. Even though the hypar has particularly efficient structural properties when used as a horizontal roof, Candela has shown, in Mexico City how exciting its form can be when used vertically as in the Iglesia de la Virgen Milagrosa. Nervi also has used vertical hypars as walls and roof in the monumental Cathedral of San Francisco (Fig. 11.30).

Thin Shell Dams

The greatest application of vertical, concrete thin shells has come not in architecture, however, but in dam construction. Although the world as a whole has so far only utilized 15 percent of the power obtainable from its natural or artificial waterfalls, the USSR has exploited 18 percent of their potential, the United States 70 percent and the three European Alpine countries (France, Switzerland, and Italy) 90 percent. Dams can be built to contain water by erecting a heavy wall of earth at the end of a valley and compacting it so as to make it watertight (Fig. 11.31). These dams resist the horizontal pressure of the water behind them by means of their weight (as the weight of a building resists the horizontal pressure of the wind) and are called *gravity dams.* They are commonly used in developing countries, where labor is abundant and inexpensive and heavy earthmoving equipment rarely available. On the other hand, where valleys are deep and their sides are formed by rocky mountains, and

11.31 GRAVITY DAM

11.32 THIN-SHELL DAM

where concrete technology is well developed, dams are often built as thin, concrete, curved surfaces, which resist the pressure of the water through their curvature (Fig. 11.32). (They may be thought of as curved roofs loaded with snow, but rotated into a vertical position, so that the snow load becomes horizontal.) Some of the Alpine dams are monumental structures reaching heights of over 1,000 feet and transmitting the thousands of tons of water pressure to the valley sides through their curvature. It may be thought ironic that such structures be called "thin," when their thickness, which increases from top to bottom, may reach ten feet. But thickness is never measured in absolute terms: what counts structurally is the *ratio* of the thickness to the radius of the curved surface, which in a dam can be as low as 1/500. It can be realized how "thin" a ten-foot-thick dam is by comparing it with the curved shell of an egg, in which the thickness to radius ratio is as much as 1/50. A dam is, relatively speaking, ten times thinner than an eggshell.

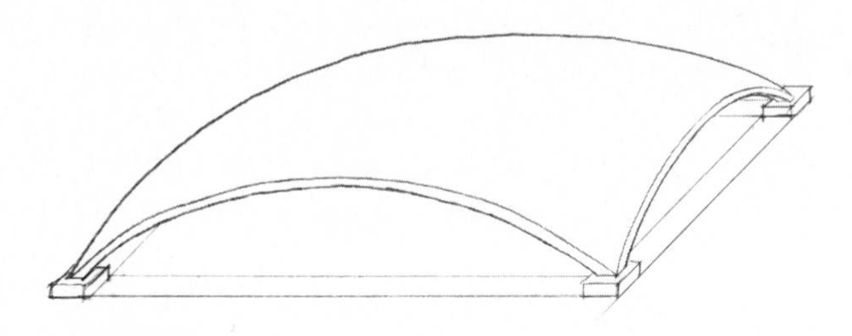

11.33 THIN-SHELL "SAIL" ROOF

One of the questions often asked of the structural engineer is whether any of the beautiful curved surfaces encountered in nature or imagined by the fertile mind of an artist could be used to build roofs or other structures. For example, one lovely, thin-shell roof in California has been built in the shape of a square-rigger sail, blown out by the wind but then turned into a horizontal position and supported on its four corners (Fig. 11.33). Although not a geometrically definable surface, it is structurally efficient. On the other hand, though an undulating surface can have a pleasant appearance, it would be quite inefficient structurally due to its tendency to fold like an accordion. We can learn a lot from nature, only if we know how to look at it with a wise and critical eye.

This chapter must end with the melancholy realization that over the last few years thin, curved shells, lovely as they may be, have not been very popular in advanced technological countries for purely economic reasons. The main obstacle to their popularity, already mentioned, is the cost of their curved formwork. Innumerable procedures have been invented and tried to reduce the cost of the formwork or to do away with it altogether. Pneumatic forms were first used in the 1940s by Wallace Neff, who sprayed concrete on them with a spraygun. Dante Bini sets the reinforcement and pours the concrete on uninflated plastic balloons, and then lifts them by air pressure. The Bini procedure, in particular, has met with success almost all over the world in the erection of round domes of large diameter (up to 300 feet) for schools, gymnasiums, and halls. Of course, balloons are naturally efficient when round. This procedure cannot be well adapted to other thin-shell shapes.

A traditional method of construction, originating in the Catalonian region of Spain, has for centuries produced all kinds of curved thin structures without ever using complex scaffolds or formwork through the ingenious use of tiles and mortar. For example, to build a dome the Catalonians start by supporting its lowest and outermost ring of flat tiles on short, cantilevered, wooden brackets and grout to this first layer a second layer of tiles by means of a rapid-setting mortar. Once this first ring is completed and the mortar has set—in less than twelve hours—workers can

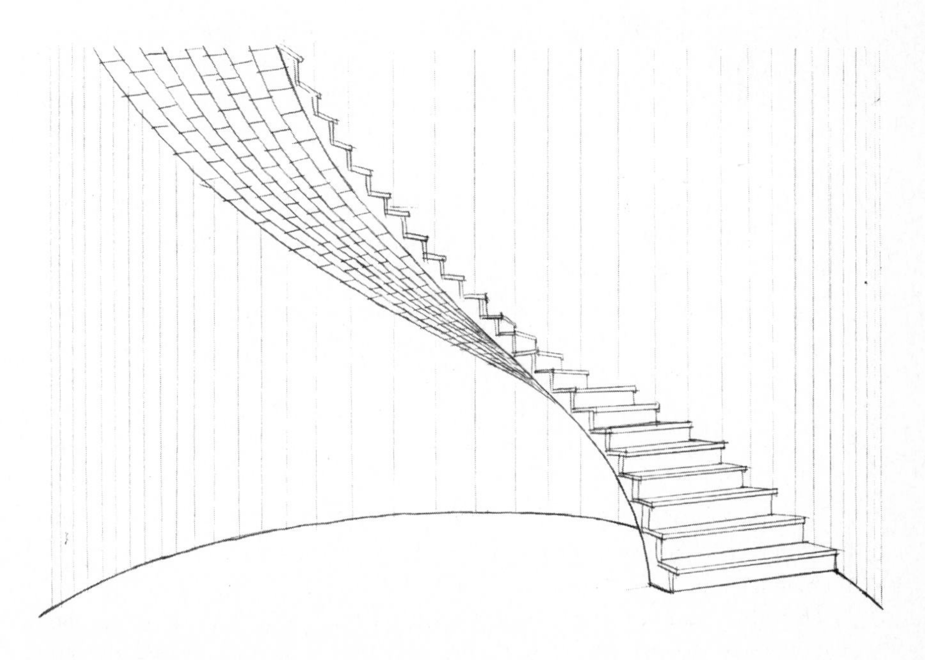

11.34 SPIRAL STAIRCASE OF THIN-SHELL TILE CONSTRUCTION

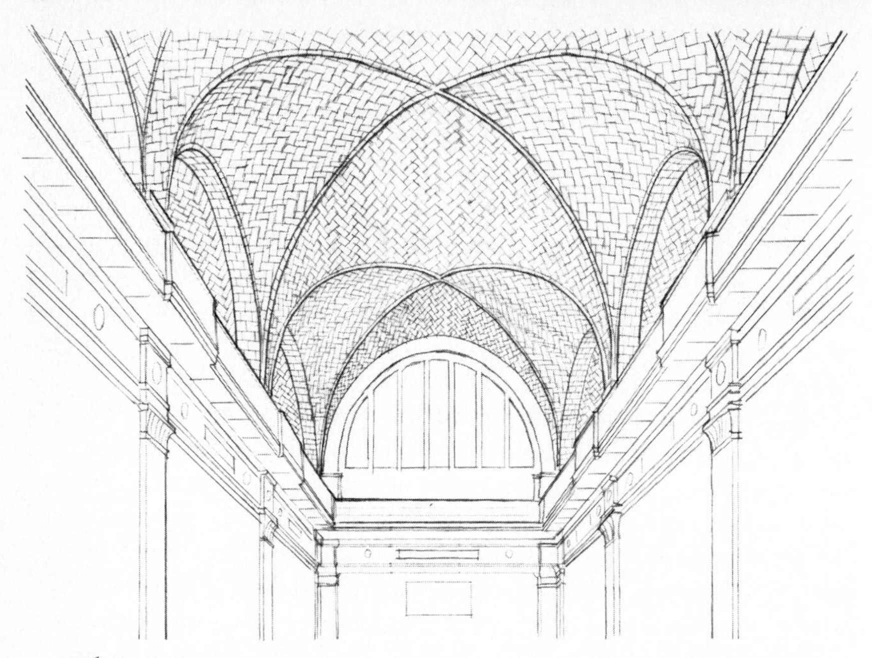

11.35 GROIN VAULTS OF THIN-SHELL TILE AT FORT McNAIR, VIRGINIA

erect the next ring by standing on the first and adding as many layers of tile as needed by the span of the dome, usually not more than three layers. By the same procedure, spiral staircases are erected around interior courtyards (Fig. 11.34), or cylindrical barrels of groined vaults built. The Guastavino Company, whose Catalonian founder introduced this method to the United States toward the end of the nineteenth century, eventually

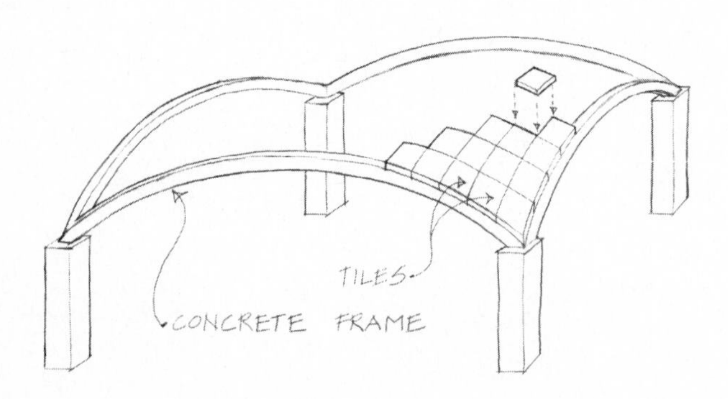

11.36 THIN SHELL OF PREFABRICATED ELEMENTS

built over 2,000 buildings in which such tile shells were used. Two of them, the dome over the crossing of the Cathedral of St. John the Divine (erected as a temporary structure while waiting for the completion of the church) and the groined vaults of the War College at Fort McNair, Virginia (Fig. 11.35), have by now been officially labelled United States landmarks. Unfortunately, the amount of labor required to set the tiles by hand has made even this procedure uneconomical. The last word on this method's use has not been said, however, since in the USSR thin shell specialists have extended the Catalonian methodology by replacing the small tiles with large, prefabricated, curved elements of prestressed concrete. These are erected without the need for any scaffold starting at one corner of a steel or concrete structural frame (Fig. 11.36).

In structures, perhaps more than in any other field of human invention, little is new under the sun, but there is always room for ingenious modifications of old ideas, as well as hope for real breakthroughs.

12 The Unfinished Cathedral

The Origin of the Gothic Cathedrals

Western culture has been blessed by eras in which the themes of political, economic, philosophical, and aesthetic life built to glorious climaxes. Such was the period that brought forth modern physics at the beginning of our century, the expansion of music and art in the seventeenth, and the explosion of the Renaissance in the fifteenth. Superficially these are revolutions, but scholars—and common sense—have shown us that the ideas of a given time have germinated for decades, even centuries. As the sudden flowering of spring requires the long preparation of winter, cultural revolutions are the consequence of cultural evolution.

A blessed convergence in twelfth-century France produced the Gothic Cathedral, one of the greatest achievements in the field of architecture. Beginning with modest, but substantial, modifications of the Romanesque style, the cathedrals of the Early Gothic period—the twelfth century—evolved triumphantly into the High Gothic structures of the thirteenth, and the impassioned churches of the Rayonnant and the Flamboyant styles of the fifteenth and sixteenth centuries. No structural style has spread as rapidly and as widely as Gothic. Twenty-five cathedrals were built between 1130 and 1230, within 100 miles of Paris (Fig. 12.1) and 80 cathedrals and 500 abbeys were built in 90 years from 1180 to 1270 under three Capetian kings: Louis VIII (1187–1226), his father, and his son. Gothic structures sprouted throughout England, Spain, Germany, and

Belgium, and the influence of French Gothic rolled outward through the Christian world under King Louis IX (Saint Louis, 1226–1270) to be blocked only in Italy, south of Milan, by the Renaissance and in Greece by the Byzantine tradition.

Why did the Gothic arise at just this time in places like the Île - de - France? Perhaps the main contributory factors were of a cultural and a political nature. Through Arabic translations the philosophy of the Greeks had, at long last, reached the center of Western culture. An entirely new atmosphere—one of freedom of inquiry, of balance between transcendent religious thought and pragmatic study of man and nature—made its appearance. Without losing his deep religiosity, the new man saw himself and the world around him as worthy of study. An understanding of the visible world became a better way of understanding the greatness of God.

The churches reflected this new spirit. The Romanesque cathedrals had been massive, dark structures, where pious men of the Middle Ages hid in fear and looked for God. The Gothic cathedrals, conversely, opened themselves to the light of the outer world, transforming it, making it unearthly. They appeared transparent and diaphanous. Their unfathomable height expressed the aspiration of humanity toward a God to be

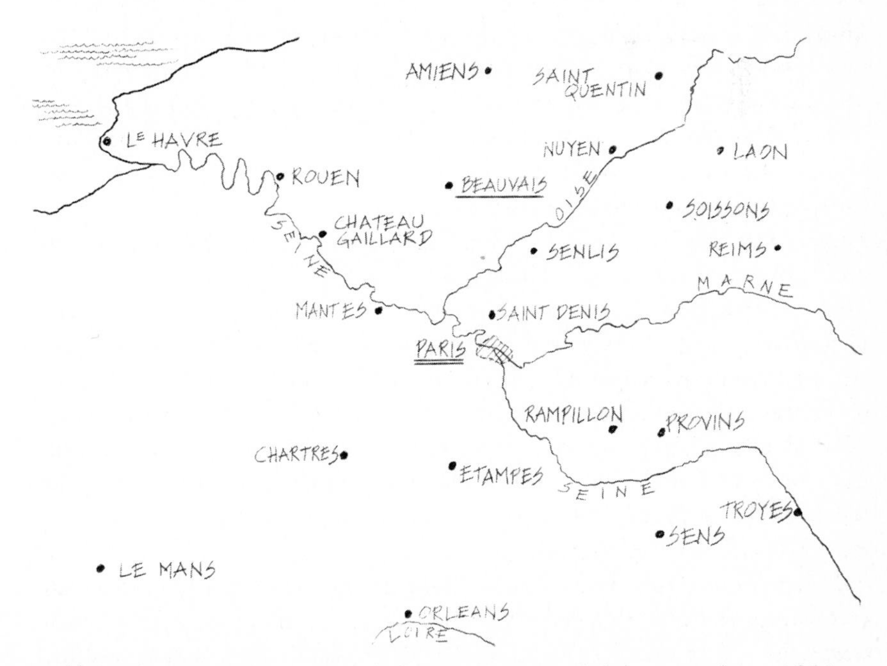

12.1 GOTHIC CATHEDRALS BUILT AROUND PARIS FROM 1130 TO 1230

12.2 THE CATHEDRAL OF CHARTRES

loved and sought in the nave by the light of day, as well as in the penumbra of the candlelit chapels (Fig. 12.2).

On the other hand, the magnificence of the statuary and the monumentality of the structure in Gothic cathedrals were signs of a new well-being, a prosperity that allowed much of the worldly goods to be spent for spiritual purposes. This luxury could only have been amassed through a new social organization. The Capetian kings of France, by subtle guile and raw power, overcame at last the predominance of the small feudal lords, and thereby concentrated power in the court, extending it over nearly the whole of France. Cities and towns flourished under the new system, commerce expanded greatly—both internally and externally—and a new type of man emerged, freed from serfdom to the local master and allowed to substitute money payments for personal services to the king. The University of Paris, second oldest in the world, opened in 1200 and the cathedral schools took over from the monasteries the responsibility for education and the spreading of new ideas.

The Gothic style is a triumph of architectural invention. But even so, and contrary to the message of their exteriors and interiors, Gothic cathedrals do not have large spans, they are not as high as some monuments built centuries earlier, and they are not "daring" from a modern structural point of view. Nor is their appearance an "honest" expression of their structural behavior. Yet one of the most structurally minded of modern engineers, Pier Luigi Nervi, considered them masterpieces. All who look upon them are awed by their apparent immense height and by the miraculous play of light and shadows through their evanescent walls and between their slender piers. The Gothic cathedral is a victory of the architect over weight and space and the purest expression of spiritual needs met by the concreteness of heavy stone. How this miracle was achieved demands a simple description of the building of a cathedral.

Gothic Spaces and Structures

One amazing feature of the Gothic cathedral is that, although its architecture evolved through centuries, one is nevertheless able to describe a "typical" cathedral. Variations from this theoretical prototype are so subtle that one can easily recognize a Gothic church in all its common components, whether it was built in the twelfth or in the twentieth century.

A cathedral, from the viewpoint of the church hierarchy, is merely the seat of a bishop, his *cathedra,* supported by the members of the cathedral's Chapter. Hence, prime movers in the construction of Gothic cathedrals were the powerful bishops of France, men who mediated between Rome and the French kings.

Bishop Milon de Nanteuil published, in 1225, a document proposing the reconstruction of the church at Beauvais, which had been destroyed by fire. He pledged at the same time, as a bona fides of his high seriousness, ten percent of his income to the enterprise and requested the Chapter to do the same. To make his request even more binding, he had obtained approval for it through a papal decree. It was the Bishop who chose the master for the cathedral, a builder whose name has vanished from history and remains known only as the First Master of Beauvais. Once these financial preliminaries had been thoroughly arranged, the construction could start, as it had been started in the two most important previous cathedrals of the High Gothic period—in Chartres in 1194 and in Bourges in 1195—and in the royal abbey of the Early Gothic period, Saint Denis in Paris, consecrated in 1140.

The plan of a Gothic cathedral has the shape of a cross (Fig. 12.3). The lower arm of the cross is represented by the wide central nave, flanked by two (inner) aisles and, often, by two additional outer aisles. The horizontal arm of the cross, called the *transept,* extends outward at right

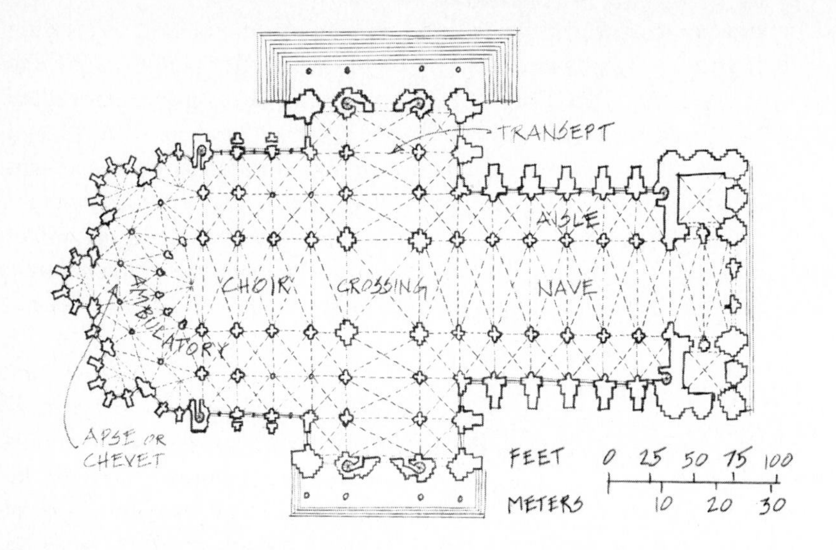

12.3 THE PLAN OF CHARTRES CATHEDRAL

angles to the nave and aisles. The main facade of the church is normally at the bottom of the nave, but the entrances at either side of the transept are often as magnificent as the main facade.* Often the point of crossing between the transept and the nave is topped by a high spire.

Beyond this *crossing*, the nave and aisles are prolonged into the choir, where the stalls for the Chapter members and the main altar are located. The upper arm of the cross, the *choir*, is closed by the apse or *chevet*, a semicircular wall pierced, usually, by radial chapels. (Additional chapels often appear along the sides of the outer aisles.) The span of the nave could reach forty-five feet, while the aisles are generally twenty to thirty feet wide. The inner aisles curve behind the altar to create the *ambulatory*, onto which open the radiating chapels of the chevet. As can be seen from this description, the plan of a Gothic cathedral does not differ much from that of most of our churches.

It is when we begin to consider the vertical structure of the cathedrals and the interior spaces they define that we meet an amazing conception, totally new from both an architectural and a constructional point of view.

To obtain a feeling of aspiration toward heaven, the Gothic masters used two architectural means: height and light. Height was not appre-

* Yet, it must be noticed that some cathedrals have no transept, like the Cathedral of Palma.

hended suddenly but in successive steps, as the eye was led to the highest point in the enclosed space, the ceiling of the chevet. The ceiling of the outer aisles was low as in the radiating chapels of the chevet; the ceiling of the inner aisles was higher and—in some cathedrals—equal to that of the ambulatory; the ceiling of the central nave rose still higher, and that of the choir and the chevet, highest. Thus in successive steps the eye is led to a point perhaps 150 or more feet above the level of the church floor (Fig. 12.4). This visual climb is emphasized dramatically by the increase in light which follows the increase in height: the outer aisles are, if not dark, shadowy; the inner aisles have large windows and greater light; the nave and, above all, the chevet are inundated with light from the tall stained-glass windows, which made the high ceilings float above the entire church. All who have visited a Gothic cathedral are familiar with this feeling of being transported toward the incorporeal ceiling of the chevet. Moreover, the colored glass of the windows transforms the light in the cathedral. It does not seem to come from the outside, but has an ethereal, other-worldly quality that separates the dematerialized interior from the reality of the outside world, while avoiding the feeling of containment and oppression typical of the Romanesque cathedrals.

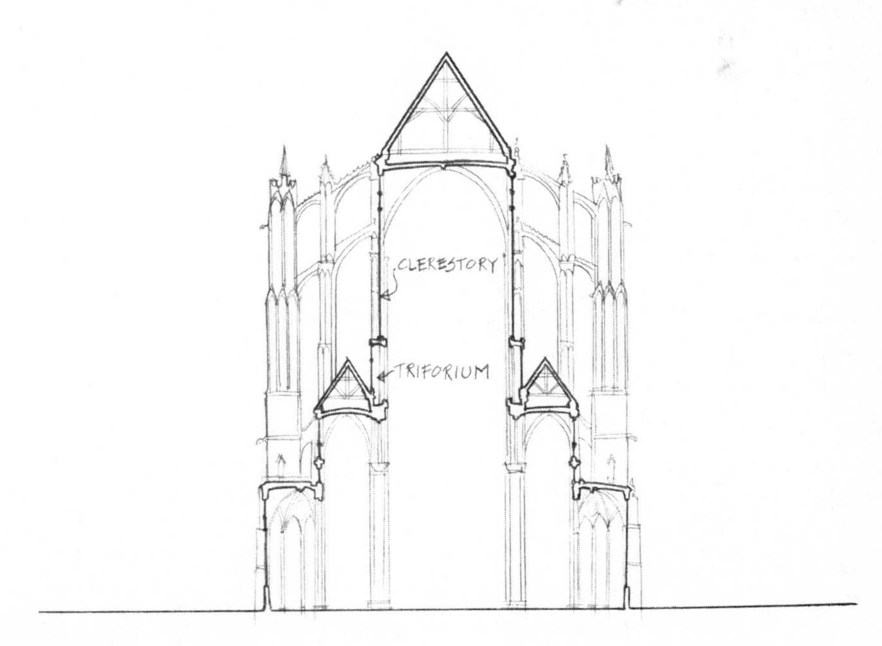

12.4 SECTION OF BEAUVAIS

The gradation of light towards the top of the church is modulated by two elements appearing in all the vertical walls into which windows are opened. The lighted areas are divided into horizontal zones: the *triforium,* a band arabesqued by carved arches, which in some cases has small window openings at its top, and the *clerestory,* the area of the long, tall windows subdivided into three or four slender compartments by means of thin stone mullions, often starting in the triforium and moving up to the clerestory in straight vertical lines as thin as pencil strokes. Thus, the light from the clerestory is supported by the darker light from the triforium, giving additional modulation to the three light steps of the outer aisles, the inner aisles, and the chevet and nave. Six light gradations leading the eye to the vertex create a tremendous effect even on a single wall, but this effect is greatly emphasized by the fact that the three main light gradations occur on three separate walls, moving into the inner space from the outer limits of the enclosure or *vessel.* The first triforium and clerestory are open in the walls of the outer aisles, the second in the wall between the outer and the inner aisles and the third in the walls between the inner aisles and the nave. The crescendo of light does not occur on a plane but on a succession of inward-moving planes that make the interior space soar, like a stepped pyramid, to the vertex of the cathedral. Height and light have produced the miracle.

We have referred vaguely, so far, to the ceiling of the various parts of the church in order to end our description with the last great feature of the Gothic interiors. While Romanesque churches are usually covered by trussed, wooden roofs or stone barrel vaults, which do not permit large window openings, the aisles, nave, transept, choir, and chevet bays of a Gothic cathedral are all covered by masonry vaults. One may understand, and justify, the entire masonry structure of the Gothic churches by analyzing their ribbed, groined, four-sided vaults. These consist, usually, of the intersection of two half-cylinders (Fig. 12.5) and, thus, allow arches to appear on all four sides of the bay they cover—a feature that obviously increases the quantity of light at the roof level.

The path of the loads carried by the ribbed, groined vaults makes the structure of the cathedral self-explanatory. The intersections of the two cylinders of a *quadripartite* (four-sided) vault constitute arched folds or groins along its two diagonals (Fig. 12.5), which, as we have seen in Chapter 11, give greater rigidity to the vault in the direction of the groins. Hence, the vault's weight tends to be channeled by the groins to its four corners, where it is supported vertically by the piers. The downward-curved ribbed groins, acting as arches, thrust out on the piers. There are

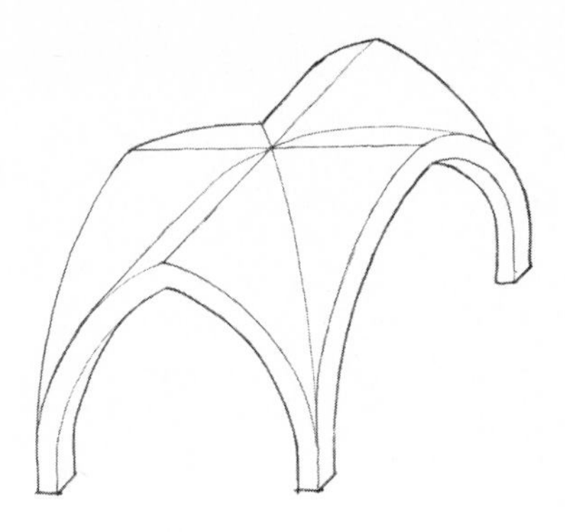

12.5 GROINED VAULTS AT BEAUVAIS

two ways of resisting these outward thrusts: by inserting tie-rods (either along the four sides of the bay or diagonally across its corners), or by supporting laterally the top of the piers by means of buttressing elements (see Chapter 9). The Gothic masters rejected the first solution—as a permanent solution—for both aesthetic and practical reasons. Tie-rods across piers would have ruined, visually, the vault-defined spaces, and their rusting (even if they were covered with lead), might, in time, have cracked the masonry in which they had to be anchored. Therefore, the first preoccupation of the masters was to reduce the magnitude of these thrusts. They had little knowledge of structural theory, but experience had shown them that pointed arches thrust out less than circular arches. The main difference between Romanesque and Gothic arches lies in the pointed shape of the latter, which, besides introducing a new aesthetic dimension, has the important consequence of reducing the arch thrusts by as much as fifty percent. (In a written statement in the year 1350 the builders of the Milan Cathedral went so far as to assert that "pointed arches do not thrust on buttresses," but they probably knew this not to be true and used it as a last line of defense against their French consultant, Mignot, who had suggested different arch proportions.) The Gothic arch is a typical example of an aesthetic feature dictated by structural requirements and may be said to be more "correct" than a circular arch in the context of large roofs. As pointed out in Chapter 11, groined vaults—even if not ribbed—are self-supporting when their thrusts are resisted, but the masters emphasized the role of the groins and of the side arches by means of ribs, which fool the eye into believing that they channel the loads into

12.6 THE CHEVET OF BEAUVAIS

the piers. By prolonging these ribs along the piers' surface without interruption—sometimes all the way to their bottom—the Gothic architects created a series of continuous lines which are a "pseudo visual path" of the loads and seem to express on the interior of the cathedral a fake "framed" structure (Fig. 12.6).

If the piers had not been so tall and thin to achieve the visual quality of the space wanted by the masters, the "frame" could indeed have supported the structure. But the aesthetic requirements, of primary importance to these supreme artists, did not allow the piers to resist the horizontal arch thrusts or, sometimes, even to support the weight of the vault's masonry, which consisted of an inner layer of finished stone topped with rubble concrete. Since the piers would bend under the action of the thrusts and, possibly, buckle under the action of the vertical loads, the groined vaults, although ribbed, *had* to be buttressed. Here again two solutions were available to the masters: internal buttresses (obtained by connecting the piers of the nave to those of the aisles by means of transverse walls, as in St. John the Divine in New York), or buttressing elements external to the church. There is no question that the first solution would have frustrated the goal of creating a single, open, high, lighted inner space. It was rejected. Instead, the vaults were buttressed by means of wall-like pillars set outside the church, which at first were attached to its walls and acted as external "shear walls" (see Chapter 7), but beginning in about 1170, were set away from the church walls and connected to them by means of inclined *flying buttresses* which support the groined vaults like slender, gigantic fingers (Fig. 12.7). The interior of the cathedral thus remained linearly pure–its space uncluttered by intersecting structural elements–while the exterior became cluttered with a magnificent forest of vertical pillars and flying buttresses.

The logic of this structural system is unchallengeable and the results superb. The refinement of the exterior elements, designed by men who were mainly guided by geometrical concepts of proportion and had little quantitative understanding of structures, is nothing less than amazing. The flying buttresses, for example, have a straight upper surface and a curved lower surface so that their almost straight axis follows the line of the vault's thrusts, while their slightly arched shape shows how they support their own dead load by arch action, without introducing tensile stresses in their masonry. In order to reduce the dimensions of the pillars, two of them rather than one were often used to resist the vaults' thrusts, connected by flying buttresses in two flights and two tiers (Fig. 12.7). Moreover, heavy *pinnacles* were added on top of the outer pillars, which with their own

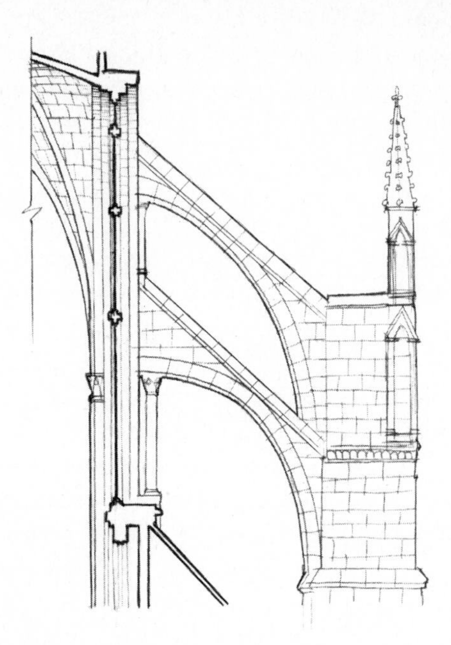

12.7 FLYING BUTTRESSES

weight pushed inward the vaults' thrusts. The more nearly vertical resulting forces acted on the pillars with an increase in compression which the masonry could well take and a reduction in bending which might have caused unwanted tensile stresses. It has been thought that even the heavy statuary, profusely spread over the outer surface of the church, was located at times so as to combine its vertical load with the horizontal vaults' thrusts and act as pinnacles.

The cathedral masters were supreme artists first and second, out of necessity, good engineers. Almost nothing is known about how they designed their structures, but we may presume from their later organization that they belonged to strictly controlled guilds and only became masters after a long apprenticeship, which must have proceeded through the steps of a stiff hierarchy. Masters consulted among themselves, but did not divulge their secrets to outsiders—not even to the commissioners of their work. One of them killed his son upon learning that he had leaked trade secrets to the bishop of the cathedral they were building. There seems to be little doubt that one of the basic methods they used to acquire knowledge was the honored process of trial and error. As time went on they became more and more daring: in the 156 years separating Chartres

(1194) and Palma (1350), the piers of the cathedrals became three-and-a-half times thinner.

Even so, the use of structural materials in the cathedrals was very conservative; their masonry is stressed to a small fraction of its capacity. On the other hand, the articulation of their structure is quite risky since it makes their stability depend on the interaction of *all* its elements. While all its components work harmoniously, a cathedral is a safe structure, but if even a minor component malfunctions, the entire frame is endangered. By pushing the complexity of the resisting elements to the limit in order to attain new aesthetic goals, the masters, unguided by sound technical knowledge, were bound to court disaster. And this brings us to the tragic story of the most beautiful cathedral of them all, Beauvais.

Saint Pierre at Beauvais

Beauvais was a lovely medieval town about forty miles north of Paris, until the Germans destroyed it during World War II. Of its fifty-five churches only three stand today. Luckily, one of them is the Cathedral of St. Pierre. St. Pierre has been called "the most famous and elongated of the French cathedrals," "the marvel of the medieval style," "its ideal," "the Parthenon of French architecture." And yet St. Pierre, after collapsing twice, was never finished. It has no tower, no nave, and no aisles. This most celebrated of French cathedrals consists of a choir, a transept, and a chevet. How did this happen?

We have seen how in 1225 Bishop Milon started the construction of St. Pierre and appointed the First Master, whom he may have known at Chartres—a cathedral also built, like St. Pierre, after an earlier church was gutted by fire. The First Master was certainly familiar with Chartres and Bourges, and he was both an original artist and a good engineer. He conceived the cathedral in the great tradition of his predecessors. Nonetheless, he contributed a number of new variants to their basic themes. To begin with, he made his cathedral more luminous by increasing the span between the piers of the central vessel up to 27 feet, so as to allow more light to come into the nave from the aisles. He also made the aisles more luminous by openings in all their walls, chapels, triforia, and clerestories (see Fig. 12.6). He elevated the vaults of the choir and the chevet—which eventually reached the unprecedented height of 157 feet 6 inches—and pierced the outer walls of the church with the lightest clerestories ever dared. His conception was grandiose, but he was also capable of devising the subtlest space modulations in order to increase the triumphal march

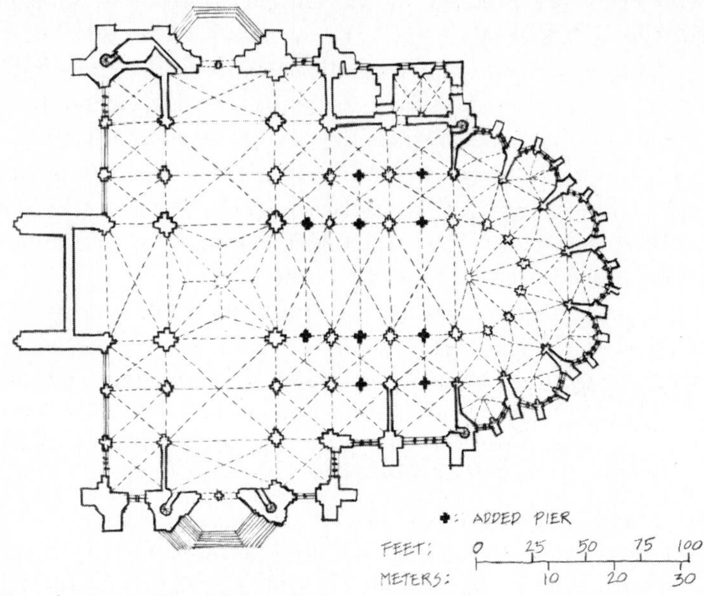

12.8 THE PLAN OF BEAUVAIS

from the church entrance toward the chevet. The spacing of the piers of the choir was not constant, but increased as one moved towards the chevet (Fig. 12.8). The radiating chapels of the chevet are relatively small and each is lit by three windows, subdivided into two narrow compartments by thin mullions. These lead to the triforium of the ambulatory, which, in turn, leads to the amazingly light triforium and clerestory of the chevet (see Fig. 12.6). Uninterrupted ribs rise from the bottom of some piers to become the ribs of the vaults. The interior of St. Pierre is rightly considered the masterpiece of High Gothic.

On the exterior the vaults are supported by two piers, connected to the main vessel of the church by two flights of flying buttresses in two tiers (see Fig. 12.7). The flying buttresses abut a pier in the wall of the church, right above heavy stones supporting enormous statues. The transepts were built in the Flamboyant style in the sixteenth century, but the church has no west facade, except a naked wall from which protrudes the small nave of the original Carolingian church, dwarfed by the Gothic chevet (Fig. 12.9).

The First Master worked at St. Pierre for twenty years. He set solid foundations and erected the chevet up to the level of the inner aisles. The execution of his masonry is technically perfect, crafted with well-cut stones and careful joints. Some historians believe that in his design (of which we have no trace) the height of the main vaults of the choir was lower than that of the present vaults. After five years of work under another anonymous master, called the Second Master, the very daring Third Master (also unknown by name) took over and finished the choir and the chevet in 1272. And then, quite without warning, the main vaults of the choir collapsed on November 29, 1284.

Why the structure stood for twelve years and then suddenly collapsed remains a mystery to this day. It is true that the masonry of the Third Master is of lower quality, but no obvious faults in either construction or design have been discovered so far, even through the use of the most advanced methods of structural analysis. Viollet-le-Duc, the great architectural historian and restorer of the nineteenth century, suggested that the slow creeping of the masonry mortar could have transferred some of the load of the walls to the piers, dislodging at the same time the heavy stones supporting the massive statues. These stones were supported by a wall pier and by two extremely slender outer columns (Fig. 12.10). According to Viollet-le-Duc, the added weight on these two columns buckled them (some are actually buckled today) allowing the heavy stones to rotate

12.9 THE WEST FACADE OF BEAUVAIS

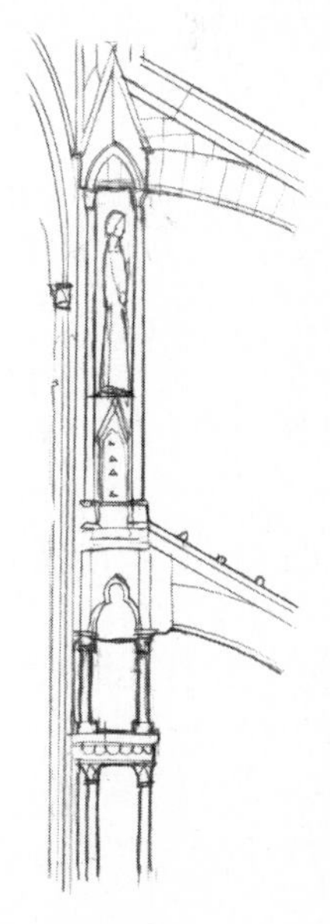

12.10 PIER WITH STATUES (VIOLLET-LE-DUC)

12.11 BRANNER'S RECONSTRUCTION OF THE ORIGINAL INTERIOR OF BEAUVAIS

outwards. This would have, in turn, weakened the connection of the upper flying buttress to the wall pier and caused its fall. Once the flying buttress fails, the entire vault system loses stability and the vaults are bound to collapse. Robert Mark has suggested, instead, that the alternate action of the wind on either side of the church overstressed the intermediate external pillar, which failed. Jean Heyman has proved that the cathedral was perfectly stable under its own dead load and attributes the collapse to a minor unknown cause.

Stephen Murray, through careful study of the reconstructed masonry, has reached the conclusion that the intermediate external pillar, which did not reach ground but was supported by an arch (en port-à-faux) (Fig. 12.4), collapsed. There have been no earthquakes recorded in the area and the foundations of the First Master do not indicate uneven settlements. In spite of all this learned research, to this day we do not know with certainty why the vaults collapsed.

The choir repairs were finished in 1337. But, unfortunately, the new unknown Fourth Master in charge of the reconstruction, apparently attributing the collapse to an excessive span between the piers of the nave and of the inner aisles, decided to cut in half these spans by erecting six intermediate piers in the main vessel of the chevet and four in the inner aisles (see Fig. 12.8). Although the interior of the church is still magnificent (see Fig. 12.6), the visual reconstruction, by Robert Branner, of its appearance in 1272 (Fig. 12.11) shows how much lighter and more daring had been the design of the First Master. The erection of the interpolated piers changed the choir vaults into sexpartite (six-sided) vaults (Fig. 12.12), thereby requiring the construction of additional external pillars and flying buttresses (in two flights and two tiers). Whether the choir could

12.12 SEXPARTITE RIBBED GROINED VAULTS AT BEAUVAIS

have been repaired without changing the interior of the First Master remains a moot question. The fact is that the vaults have been standing ever since 1337, proving that the engineering judgment of the Fourth Master was correct, if possibly overconservative.

The Hundred Year War and the English occupation of the Île-de-France interrupted work on the church for the next 150 years. In 1500, Martin Cambiges, the Fifth Master of Beauvais and the first known by name, was given charge of the construction of the transepts. When he died in 1532, the Sixth Master, Jean Vast, finished them. In 1544 discussions began about the construction of a tower over the crossing. There were partisans of a wooden tower and partisans of a stone tower. Outside masters were consulted. Finally, in 1558, the unwise decision was made to build a stone tower, which Jean Vast started in 1564 and finished in 1569. It reached the incredible height of 502 feet (Fig. 12.13) and the sight of it alarmed more than enchanted the members of the Chapter from the moment it was completed—and with sound reason. Two years later, the four central piers of the crossing, which supported the tower, were found to be out of plumb by two and four inches on the chevet side, and by six and eleven inches on the unbuilt nave side. These last two piers, unbuttressed by the nave, were giving in under the enormous added weight of the tower. It was suggested that the first two bays of the nave be built immediately, but the Chapter vacillated, sought additional expert advice, and only decided to proceed on April 17, 1573. Thirteen days later, on April 30 (Ascension Day), the tower collapsed just after the cathedral had been vacated by clergy and parishioners, who had started on a procession. The three men left behind were, miraculously, unscathed and since 1577 on Ascension Day a celebration is held in the cathedral to remind the faithful of this miracle.

The horror created by the failure was such that nobody dared dismantle what remained of the ruins until a criminal was induced to attempt it with the promise of his life if he succeeded. Legend has it that he accepted the offer and had just started when he slipped, and would have fallen to the ground had he not grabbed a rope hanging from one of the roof beams. "Thus," said a French historian, "the rope that should have been the end of this wretch became his salvation."

The fallen tower was never rebuilt. Money set aside for the construction of the nave had already been spent and the original impulse for the construction of "the greatest of all Gothic cathedrals" had been dissipated. Moreover, as the French historian Desjardin wrote: "This was not the time to build cathedrals anymore. The schools for masters, sculptors, glaziers,

12.13 THE SPIRE OVER THE CROSSING AT BEAUVAIS

12.14 BEAUVAIS EXTERIOR

and painters, which had been inspired by their construction, were dying all over the place." In 1605, the Chapter took the decision to end the erection of St. Pierre, to leave it as a choir, a chevet, and a transept, without a nave (Fig. 12.14). Thus it happened that while the greatest church of Christendom, Saint Peter's in Rome, whose construction lasted 181 years (from 1445 to 1626) was being triumphantly finished, the greatest Gothic cathedral was dying.

Failure of the vaults at Beauvais in 1284 may have been due to a minor unknown fault in design or construction, but the tower collapse in 1573 shows a serious ignorance of structural principles and indicates that intuitive understanding of these principles had decreased rather than improved during the 350 years since the start of St. Pierre's construction. This may have been due to the decline of the trade guilds through a lack of transmission of their secrets from generation to generation. The outsiders' advice given the Chapter, leading to their choice of a stone tower, indicates that this lack of structural knowledge was widespread and not limited to the Beauvais guilds. Once more, human factors—political and economic—lay at the bottom of a situation which was to have the gravest consequences in the field of an architectural and structural endeavor. It is ironic that only 200 years later the first scientific approach to structural problems was going to be undertaken in the same Île-de-France, leading directly to the recent victories in our fight against gravity and wind.

13 Domes

The Largest Roofs in the World

One has only to mention their names: the Pantheon, Hagia Sophia, Santa Maria del Fiore, Saint Peter's, Ahmet's Mosque, Saint Paul's, and then the Astrodome, the Louisiana Superdome, the Pontiac Stadium. Domes. Images of majesty and communion. Spaces enclosing and protecting 20,000, 40,000, 80,000 people who respond in unison to some of the basic emotions of the human spirit. Monuments defining a city and visible for miles around. Mountains and caves. Climaxes of the arts of an era, where architecture, sculpture, painting and mosaics conspire to create a unique atmosphere, triumphant or subdued. Victories of technology over gravity and wind.

These structures are the most perfect examples of spatial geometry, whether realized in stone, brick, concrete, or steel. No masterbuilder's achievement has attracted humanity as has this most perfect of all shapes. Is it because of its Platonic purity? Is it because of the separation it erects between the outer, undefined space and its well-defined enclosure? Is it because of the pious or joyous feelings of fraternity it elicits from the participants in its rituals? Or because of the equality created among the throngs? Perhaps the dome is the nearest materialization of heaven, the only man-made representation of the sky, and this is why a dome seems to protect us like the sky of a clear night, embracing us and our smallness and solitude.

The perfect dome has no scale, no frame of reference. When small it may feel limitless, when large it may embrace us like a room. The image of Christ, the Pancreator, looms as immense from the dome of a small Byzantine church as from the inner surface of the largest cupola in the world. A dome is even more unfathomable in a mosque, where the absence of human images makes more difficult the measure of its dimension. Its abstract designs make the inner surface of a Moslem dome even more mysterious, while its brilliant colored tiles, flaming in the sun, make its outer surface aggressively more triumphant. Growing from its modest archeological precursors into the superb vaults of Rome, symbols of world empire, the dome reached by a quantum leap the enormous dimensions of the Pantheon and spread all over the civilized world. Did this miracle happen because of the discovery of concrete and the availability of the water-resistant pozzolana or was it due to aspiration towards eternity? These are a few of the obviously unanswerable questions elicited by the dome, the king of all roofs and the Mecca of all believers, the sky of theatre lovers and the egalitarian roof of sports fans. The dome is the greatest architectural and structural achievement of mankind in over 2,000 years of spiritual and technological development.

The Dome as Structure

Whatever significance one may give to the dome, its structural behavior must be understood before its appearance on the architectural stage may be appreciated. Let us then ignore the minor differences of shape assumed by the dome over its historical development and think of it as a perfect half sphere of a thickness small relative to its span. Whether supported over the crossing of a church or directly on the ground, the dome must carry its own weight and the weight of the live load, including the pressure and suction of the wind and, in northern climates, the weight of snow. It is obvious that these loads must be channelled to the ground, and, familiar with the behavior of arches, we intuitively realize that the dome does it along its curved vertical lines or *meridians*. The dome reminds us of a series of identical arches set around a circular base and meeting at their top, where they have a common keystone (Fig. 13.1). Our intuition is correct, since the loads accumulate from top to bottom along the vertical meridians, which become more and more compressed as they approach the dome's support. What might surprise us upon a more careful inspection of the dome is the substantial difference between its small thickness and that of arches spanning the same distance. Another

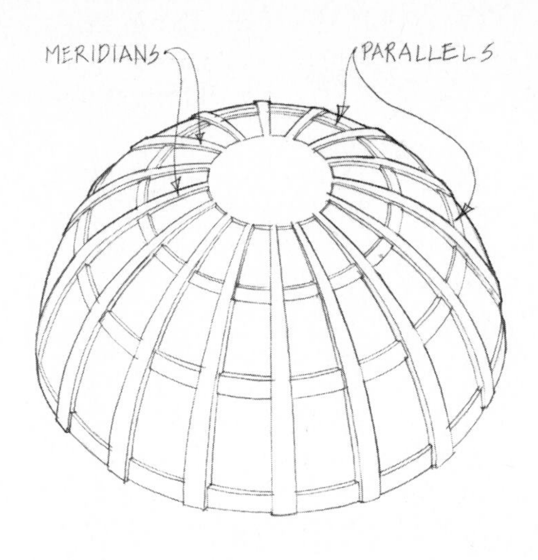

13.1 THE MERIDIANS AND PARALLELS OF A DOME

cause of puzzlement might arise if we remember that arches thrust outward and require buttresses or tie-rods to prevent their opening-up. Since the dome has no apparent tie-rods or buttresses, it soon becomes obvious that the dome is not just a series of arches set on a circle, one next to the other.

What makes the dome behave differently is the fact that the hypothetical arches it consists of are joined together along the vertical sections of the dome, making it a *monolithic* structure. Its reduced thickness and the disappearance of buttresses and tie-rods are due entirely to its monolithicity. The reduction in dome thickness as compared to arches is quite dramatic. While the thickness of a concrete arch varies between 1/20 and 1/30 of its radius, the thickness of a concrete spherical dome can be as small as 1/200 or 1/300 of its radius.

The continuity of the dome's surface allows such reduction in thickness by introducing an action along its horizontal sections or *parallels* (Fig. 13.1) that prevents the arched meridians from opening up. The parallels behave like the hoops of a barrel molding the staves, making them into a single surface. Figure 13.2 shows the (grossly exaggerated) deformed shape of a dome under load and makes it clear that the dome tends both to come down at its top and to open up at its bottom. The

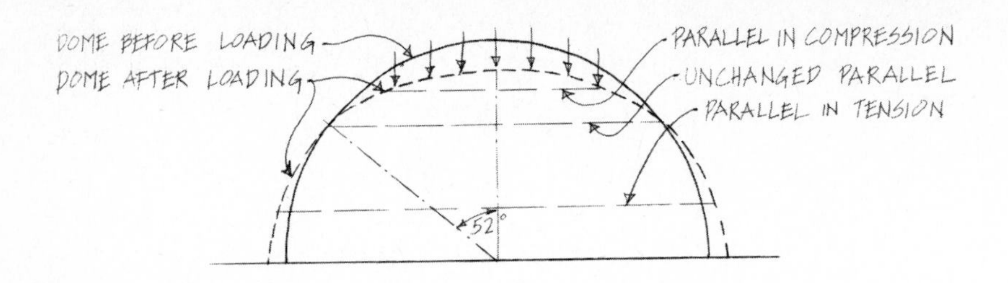

13.2 DOME DEFORMED BY VERTICAL LOADS

points on the upper part of its surface move inward, while those in its lower part move outward. These motions, which take place freely in an arch, are susbstantially prevented in the dome by the horizontal hoops. In the upper part the parallels become compressed in resisting the inward motions, which would reduce their radius, while in the lower part they become tensed in resisting the outward motions, which would increase their radius. While in an arch these deformations under load occur mostly by changing the shape of the arch, that is, by bending it, in the dome these deformations are extremely small, because, as we have seen in Chapter 4, the deformations due to compression or tension are minute as compared to those due to bending. The prevention of the meridional deformations of the dome by the compressive or tensile hoop action of its parallels has two consequences. First, it makes the dome much stiffer and prevents the buckling of the compressed meridians. Thus the dome can be made thinner without endangering stability. Secondly, it prevents the opening-up of the dome at its bottom, doing away with the need for external buttresses or internal tie-rods. Actually, the bottom parallels in tension are the circular tie-rod that prevents the opening-up of the dome.

The stiffness obtained by the combined action of the meridians, carrying the loads down, and the parallels, preventing large deformations by hoop action, is amazingly high. A reinforced concrete dome spanning 100 feet and only two or three inches thick will deflect at its top by not more than one-tenth of an inch. This deflection is only 1/12,000 of its span. In comparison the deflection of a beam is thirty-three times larger or 1/360 of its span. Moreover, the opening-up of a dome at its lowest parallel is even smaller than its deflection at the top.

Figure 13.2 shows that the upper dome parallels tend to shrink under load, while the lower ones tend to elongate. At some point there must be a parallel that neither shrinks nor elongates. It can be proved that under dead load this parallel makes an angle of about 52° with the vertical axis of the dome. All the parallels above it are in compression; all those below in tension. This behavior of the parallels of a dome was not well understood by the dome builders of the past, and without exception the domes of antiquity as well as those of the Renaissance developed vertical cracks at their bottom due to the low tensile resistance of the masonry used in their construction. While hoops of wood beams (at times reinforced with iron bars) were introduced in some Renaissance domes to prevent these vertical cracks, it was not until the bases of the domes were circled with hoops of steel chain that the cracks were prevented from opening and stability achieved.

A last basic property of thin domes must be mentioned before moving on to a description of some of their most famous examples. The reader may remember that the arch can be intuitively considered as the compressive analog of the tensile cable. It was referred to as the antifunicular shape acquired by a cable under given loads, when its funicular shape is "frozen" and flipped over (see Chapter 9). When the loads on the cable are many and closely spaced, the shape of the cable in tension becomes a curve and its corresponding compressive shape is that of an arch. It is important to notice that since a cable changes shape depending on the number, value and location of the loads on it, there is only *one* funicular shape the cable can assume under a *given* set of loads. Correspondingly, there is only one purely compressive arch shape for a given set of loads. But while a cable is flexible and changes shape when the loads on it change, an arch is rigid and cannot do so. It supports any *new* set of loads by a combination of compression along its curved shape and of bending. In other words, it acts partly as an arch and partly as a beam, and it is the beam action that causes its relatively large deformations. Here is where the dome presents an additional advantage with respect to the arch. Because it consists of a monolithic curved piece of material, it can also prevent the *sliding* of one meridian "arch" with respect to the next, by developing the antisliding stress we called shear in Chapter 5. This additional mechanism can be shown to allow the dome to carry not just one, but *any* kind of load without changing shape and without developing bending-stresses, provided the load be smooth as most loads on domes are. If one remembers how inefficient bending-stresses are, since they use properly only the material away from the neutral axis, one realizes that in addition to all of its other characteristics the dome is an extremely efficient structure in terms of the use of materials. No wonder that domes,

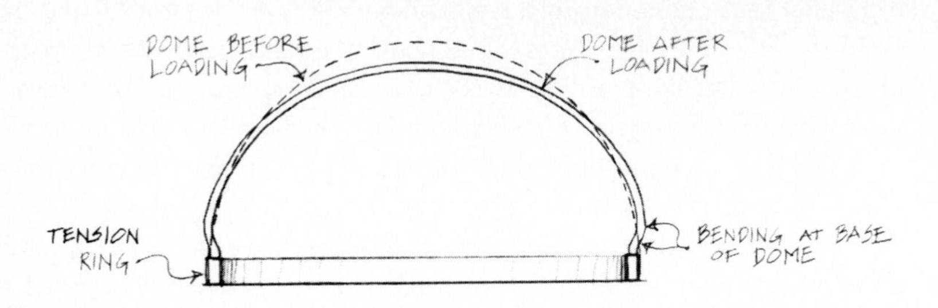

13.3 BENDING AT DOME BOTTOM DUE TO BASE RING

thin as they usually are, are stiffer and stronger than almost any other structure devised by man.

Most modern domes are built of reinforced concrete, and the required tensile resistance of their lower parallels is obtained by means of reinforcing bars located along these parallels. These bars are of larger size than those toward the top of the dome, but one must not forget that reinforcing bars are required in concrete structures not only to absorb tension but also to create a cage of steel that keeps the concrete together. Therefore, bars are always set in the concrete of a dome both in the direction of the parallels and that of the meridians.

Minute as the deformations of a dome may be, it would not be practical to allow its bottom to move in and out depending on the amount of load on it. Most domes are stiffened at their bottom by a strong ring, which to all practical purposes restrains the motion there. Figure 13.3 shows how the bottom ring, by preventing the opening-up of the dome under load, necessarily bends the dome surface in its neighborhood and introduces in it a minor amount of bending stresses. These are "snuffed out" by the hoop action of the parallels, which, thanks to their stiffness in tension, do not allow the inefficient bending stresses to penetrate into the shell of the dome. Usually not more than five percent of the dome surface develops bending stresses near its support, while the rest develops only compression and tension.

The Pantheon

The largest dome of antiquity, dedicated to all the gods, was built by the Romans under the emperor Hadrian in A.D. 123 and stands in all its glory to this day. The Pantheon (Fig. 13.4), a triumph of concrete architecture, could only be conceived and built after the discovery of

13.4 INTERIOR OF THE PANTHEON

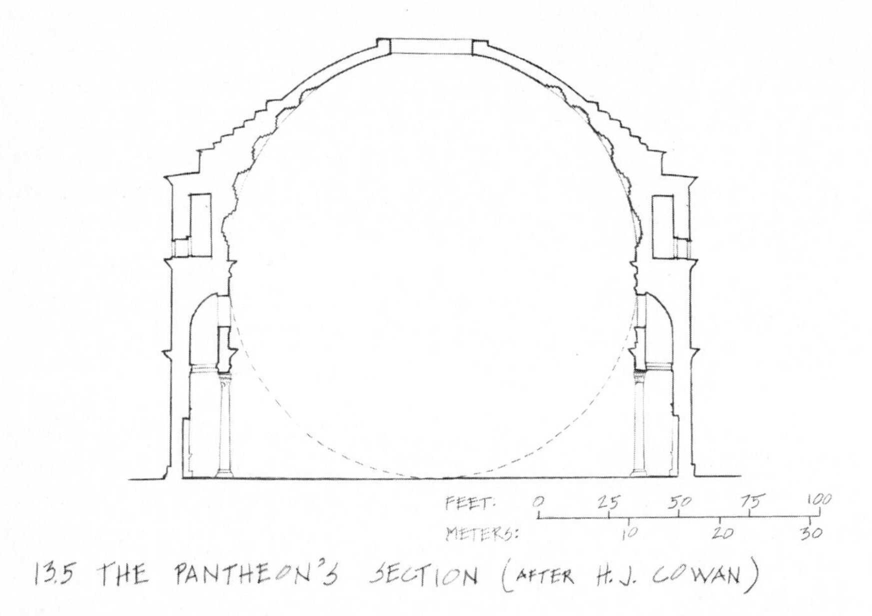

13.5 THE PANTHEON'S SECTION (AFTER H. J. COWAN)

pozzolana concrete by the Romans, who were first to erect large monolithic structures and overcome the difficulties of large spans. The dome of the Pantheon is a half-sphere erected over a circular wall of concrete and lightened by a series of decorative coffers on its interior spherical surface. Its exterior surface (Fig. 13.5), more in the shape of a cone, is responsible for the increase of thickness of the dome towards its bottom. Other Roman domes had cracked under the tension in the lower hoops and Roman architects well knew the necessity of increasing a dome thickness in the area of its support. But the thickness towards the bottom of the dome was so unnecessarily large in the Pantheon that great chambers were scooped out of it in order to reduce the dome's weight. As very dry and well-compacted concrete (typical of Roman craftsmanship) was being poured on the dome's scaffold in horizontal layers from top to bottom, the builders introduced lighter aggregates like pumice in the concrete of the upper part of the Pantheon and inserted in the top concrete empty clay amphorae, which further reduced its weight. The Romans also understood the hoop action of the dome's parallels. In order to avoid the pouring of concrete on a horizontal scaffold at the crown of the dome, they often left a circular opening or "eye" there (see Fig. 13.4). The rim of this opening was built of hard-burnt bricks, well cemented by excellent mortar, since it had to resist heavy compression in acting as the common keystone of all the meridional arches of the dome.

The dimensions of the Pantheon's dome are extraordinary both in geometry and thickness. It spans 142 feet internally, has a minimum thickness of about two feet at the rim of its top opening and a maximum of twenty-three feet at its bottom. This bottom thickness is so large that the tensile hoop stresses in it are well below the resistance of the concrete, and only a few minor cracks have appeared in the weight-reducing chambers during the eighteen centuries of its existence. The Pantheon was unsurpassed in span for 1,300 years, until the octagonal dome of the Cathedral of France out-did it by only seven feet in its maximum span.

Santa Maria del Fiore

In 1417 the "Comune" of Florence, finally in possession of a cathedral fully representative of its culture and glory, decided to erect a dome over its crossing. Two basic limitations had to be accepted in the conception of this work. First, the drum on which it was going to be supported had an octagonal shape and was surrounded on three sides by octagonal half-domes (see Fig. 13.7). Secondly, the brick model of the complete church, erected in 1367, showed the corner ribs of the proposed dome to have a particular profile, called *"a quinto acuto,"* consisting of pointed circular arches with a radius equal to four-fifths of the crossing's span (Fig. 13.6). The octagonal half-domes around the drum also had this particular profile. Both structurally and aesthetically it was inconceivable that the great dome should not have an octagonal shape or present a different profile. The all-powerful construction guilds of Florence under

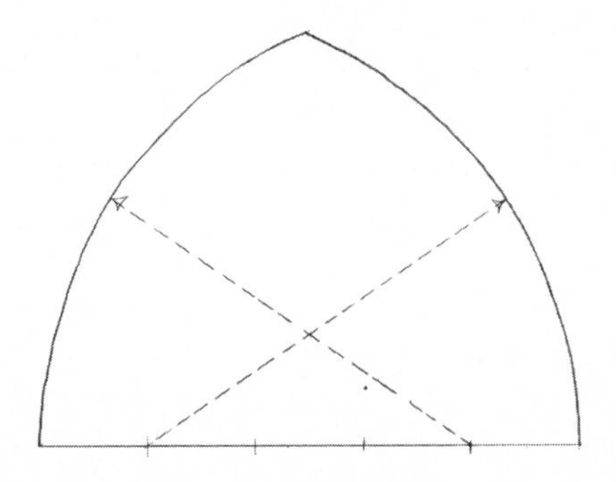

13.6 ARCH "A QUINTO ACUTO"

the direction of the Opera del Duomo (the board supervising the erection of the church) had sworn to adhere to the 1367 model, although nobody had yet made serious proposals concerning the execution of the project. In fact, it was known that as far back as 1394 the experts of the Opera del Duomo had expressed the opinion that the construction of the dome was "so big and in such a state that it could not be completed" and that "it had been naive of the earlier masters and whoever else had deliberated on the matter to have believed it could be done." The idea of a scaffold or centering capable of supporting such a monumental structure during its construction had always appeared outrageous to the experts of the time, indeed to all who had studied the problem since the Cathedral was started in 1294 by Arnolfo di Cambio and continued by Francesco Talenti in 1357. Even after the completion in 1413 of the octagonal drum on which the dome was to be supported—although the drum had been built fourteen feet thick as a matter of wise precaution—the cost of the centering alone had been a deterrent to the project. Notwithstanding these inauspicious beginnings, a competition for the construction of the dome was called in 1418. Some of the greatest architects of the time participated in it, but no winner was chosen.

In 1419 Filippo Brunelleschi, a member of the Silk Guild trained as a goldsmith, painter, and sculptor, made a revolutionary proposal to the board: the dome or "cupola" could be built without a wooden centering. He submitted a brick model to prove it. The board took him seriously enough to pay for the model, even though no large dome had ever been built without an interior temporary support.

This was a conservative time. Not far away the Sienese had renounced the grandiose concept of making their striped cathedral into the transept of a much larger church. Tuscany had been invaded and almost entirely conquered by the Viscontis of Milan. The art of building was in the hands of specialized guilds opposed both to new ideas and to the leadership of a single man. But Filippo Brunelleschi was not easily beaten. Although only twenty-four years old, he was already well known in his town as a superb sculptor who had placed second in the competition for the doors of the Baptistery (the so-called Doors to Paradise) and had proudly refused to collaborate with the winner, the great Ghiberti, in their execution. Son of an illustrious magistrate, Brunelleschi was a humanist at heart; a great mathematician particularly enamoured of geometry, a mechanic and inventor of tools and clocks, a scholar who had dedicated years to the study of the *Divine Comedy* in order to grasp the many meanings of the great poem. Having in his opinion failed as a sculptor,

13.7 THE DOME OF SANTA MARIA DEL FIORE IN FLORENCE

he had travelled at his own expense to Rome to study its monuments and had come back to Florence fired with a sense of architectural mission. This time he did not fail. Over a period of only twenty-seven years Brunelleschi created a new architecture, crowned by the completion of Santa Maria del Fiore which, together with the design and erection of the Pazzi Chapel, the Foundlings Hospital, the Old Sacristy of San Lorenzo, the new Church of San Lorenzo and that of Santo Spirito, as well as numerous palaces, transformed Florence from a medieval town into the capital of the Italian Renaissance.

By 1420 the board of the Opera named Brunelleschi and Ghiberti co-architects of the dome of Santa Maria del Fiore, and Battista d'Antonio master of the works, or as we would call him today, engineer of record. Brunelleschi accepted the partnership, then, guilefully feigning illness, showed Ghiberti to be incapable of proceeding alone and took complete charge of the project. He is referred to in written specifications for the job as "the inventor."

Filippo meant business. He took residence in a house at the foot of the dome. He chose the clay and the dimensions of the large bricks for the masonry and supervised their burning. He determined from which quarries the stone and the marble for the dome would come, established kitchens inside the church so that the workers would not waste time going out for food, and erected scaffolds and banisters to make the masons' work less dangerous and psychologically more comfortable. He refused to pay dues to the stone masons and woodworkers guild in order not to be taken for just another member. He trained crews to substitute for them when they struck. All in all he spent the next sixteen years of his life bringing to fruition, day by day, his revolutionary idea. In 1436 the dome was finished (Fig. 13.7). Dominating the city of Florence and the Tuscan countryside, it moved Alberti, the first historian of architecture, to write, "Who could be so hard or so invidious not to praise Pippo the architect in seeing here a structure so large, erected above the heavens, so wide as to cover with its shadow all the people of Tuscany, built without any help of centering or large amounts of wood, an invention which, if I am a good judge, as it was incredible in our own time that it could be done, it was not conceived or known by the ancients."

The "incredible invention of Pippo" had many components, the most important of which was its double-masonry dome (Fig. 13.8). This double-dome consisted of a thick inner octagonal shell connected by meridional arched ribs to a thinner outer shell. The inner dome was thus protected from the weather, and the exterior of the cupola given a more

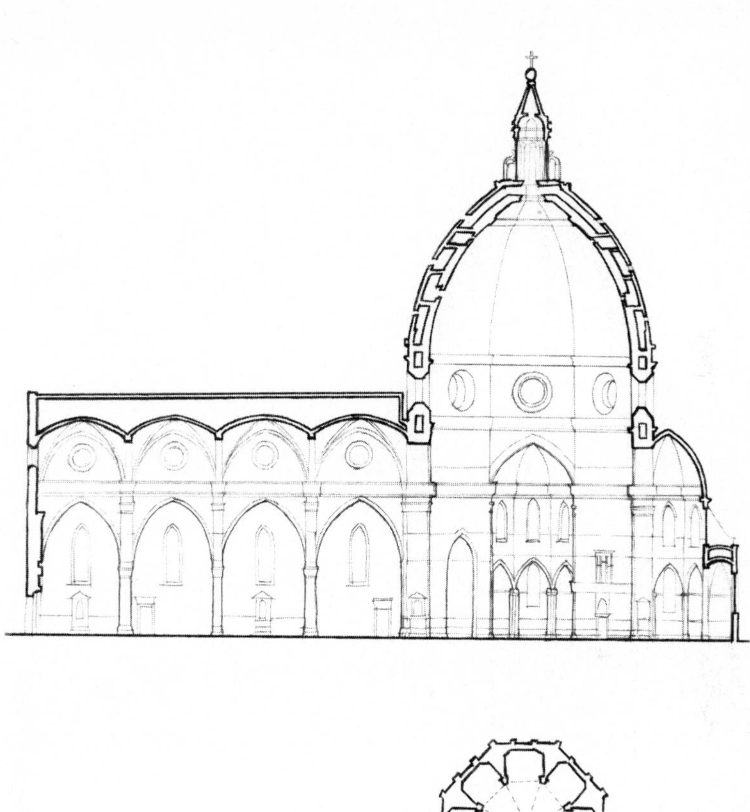

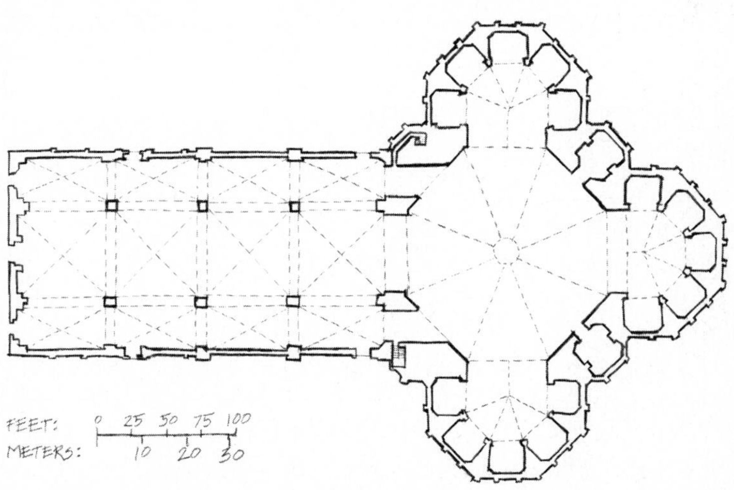

13.8 SECTION AND PLAN OF FLORENCE CATHEDRAL

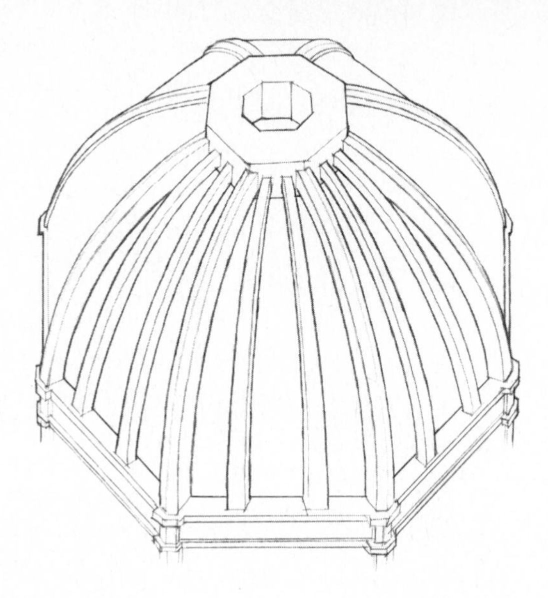

13.9 SKELETON OF THE DOME OF FLORENCE CATHEDRAL

majestic shape. Two domes increased the resistance of the entire structure and permitted the inspection of both domes from stairs and galleries meandering through the space between them. The most essential structural component of the domes consisted of its hoops—six horizontal rings of sandstone reinforced on their outer surface by iron chains, which would prevent the bursting of the domes under the enormous tensile forces in their parallels. While the Pantheon owed its strength to the brute weight of its thick masonry, the cupola consisted of two relatively thin, light domes, which relied on the strength of their stone-and-iron hoops to prevent collapse at any stage of construction.

The skeleton of the cupola (Fig. 13.9) consists of the eight corner ribs in the shape *a quinto acuto* and of two additional smaller ribs located between the corner ribs, on each side of the octagon. The corner ribs are fourteen feet wide, the intermediate ribs eight feet wide. Both are deep enough to connect the inner to the outer dome but decrease in depth, although not in width, as they grow from the base octagon to the top octagon. The top octagon acts as the keystone for all the twenty-four ribs and supports the magnificent lantern of hollow marble, then a conical roof that in turn holds the golden ball and its cross. The inner dome is seven feet thick at the bottom and five at the top; the outer is two and one-half feet at the bottom and one and one-fourth feet at the top. Both domes consist of eight cylindrical faces that curve inward toward the axis of the dome but have straight horizontal sections throughout their height. The iron-reinforced hoops of sandstone, embedded in the brick masonry of both domes at equal vertical intervals, have fulfilled their role as Brunelleschi conceived it. They prevent the bursting of the cupola and the appearance of those cracks that have required later remedies in all the large domes of the past, including Hagia Sophia and Saint Peter's.* One must not think of the cupola as of a skeleton of meridional ribs and horizontal hoops supporting the thin surfaces of the inner and outer domes. Although most of the enormous weight of the cupola is carried to its octagonal base by the wide ribs, one must realize that the twenty-four ribs and the two domes reached the same height simultaneously during construction, and this is the key to the last component of "Pippo's incredible invention," the erection of the largest dome ever built without the help of a centering.

We owe to the meticulous and untiring efforts of Rowland J. Mainstone the explanation of this miracle. The structure of the cupola can be inspected through five galleries, which run around it between its two domes—one at the base and one at the top of the base octagon, two between the base and the top octagons, and one in the depth of the upper octagon. A walk through these galleries showed Mainstone clearly that above the third gallery each corner rib is connected to the two intermediate ribs right and left of it by nine evenly spaced sets of horizontal arches and that these arches, together with the portion of the outer dome between the intermediate ribs, constitute nine concentric horizontal circles (Fig. 13.10). Below the third gallery the inner dome is so thick that a fairly thick circular ring can be drawn entirely inside it. Mainstone thus proved that although the shape of the cupola is octagonal, it contains in its interior a number of circular rings (Fig. 13.10) and hence works structurally as a circular dome! As Brunelleschi knew from his studies of the Pantheon and other Roman domes, a circular dome is stable at all times during its construction because the uppermost completed ring acts as keystone for its meridional arches and prevents them from falling inward. Hence the circular dome stands without need of inner support during its erection, provided its tendency to burst at the parallels be prevented. And this Brunelleschi had achieved by means of the iron-sandstone hoops. One may add that the profile *a quinto acuto* is so

* Cracks have been recently discovered along the main ribs of the dome, but they are not recent, as shown by the frescoes that were painted around them originally.

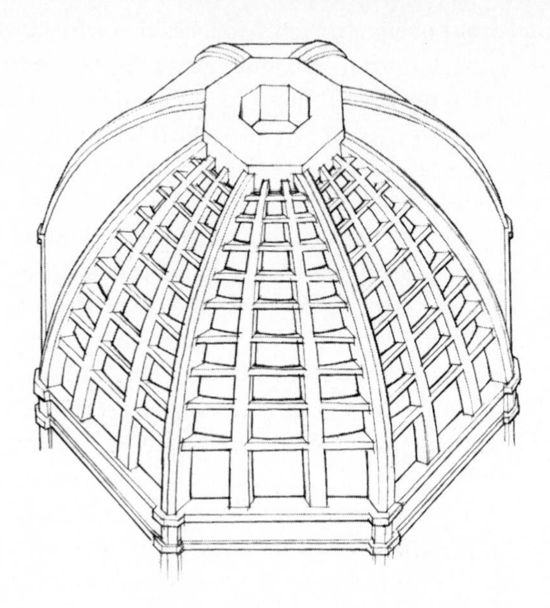

13.10 THE NINE HORIZONTAL CIRCLES IN THE FLORENCE DOME (AFTER R.J. MAINSTONE)

pointed that its tendency to burst is about half as great as that of a shallower spherical dome. This could not have been known to the board, which most probably chose the ribs' shape for pure aesthetic reasons, but it was certainly a contributing factor to the success of the enterprise.

One more detail needed to fall into place before Brunelleschi's dome could be made to work. During construction the uppermost ring of a circular dome can act as keystone for all its meridians only when it is a *complete* ring capable of resisting compression. Unfortunately during construction the uppermost ring cannot be built instantaneously, and while it is incomplete and open, it cannot support all the meridians. Brunelleschi solved this conundrum by the simple expedient of connecting the incomplete uppermost ring under construction to the lower complete rings of the domes. He simply laid the bricks of the uppermost rings flat over those of the previously built rings but interrupted these layers of flat brick by inserting vertical bricks between them every three feet. The vertically laid bricks follow a spiral curve along the surface of the domes, each tying the upper masonry to three of the lower layers, creating a herringbone pattern in the masonry (Fig. 13.11). As the time needed for the mortar to set was shorter than that required to complete a new layer of bricks, each incomplete layer under construction was always keyed to three lower layers and sustained the inward push of the meridians leaning on it. Brunelleschi also understood that for the brick masonry to work as does the continuous concrete of a dome, the bricks should not be laid horizontally but at right angles to the dome surface, that is, with an inward inclination, increasing together with the height of the dome. The herringbone spirals of vertical bricks had the additional purpose of preventing the sliding of the flat bricks on their inclined surfaces by tying them to the masonry that had already set. An inspection of the lie of the bricks of both the outer and the inner dome shows how Brunelleschi had carefully studied the influence of these construction details on the strength of the cupola and why he could so forcefully assert to the supervising board that the inner and outer domes could be built without centering and that the cupola would neither burst nor fall inward under its own weight.

Brunelleschi saw the Duomo completed and his conception proven uncannily correct. He then designed its ethereal lantern of hollow marble and the special tackle required for its erection. But he was not allowed to enjoy the completion of his labors. He died in 1446 when the lantern lay at the foot of the cupola ready to be lifted to the top of the upper octagon. His body lies inside the Cathedral next to a plaque expressing the admiration of the Florentine people for one of its greatest sons.

In the hilly countryside around Florence many hamlets and localities are called Apparita or Apparenza, that is, Apparition or Appearance. At these spots the traveller will suddenly see, as the road turns, the profile of the cupola, first just its lantern, then more and more of its unique profile . Thus he will know that he is approaching the city. But in Italian the word apparition is used almost exclusively in connection with the

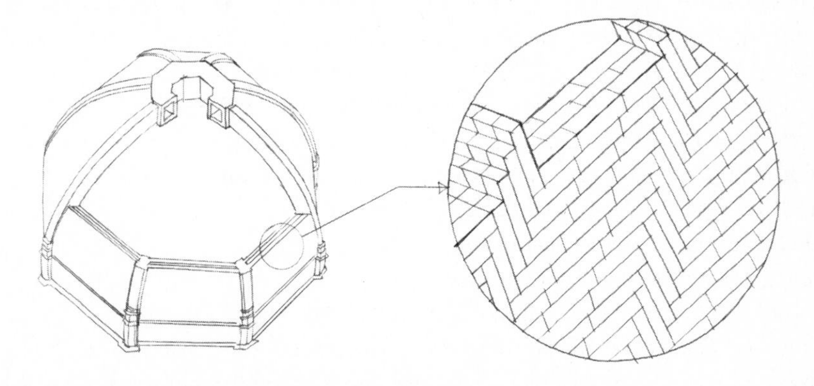

13.11 THE HERRINGBONE MASONRY PATTERN

miraculous apparition of the angel to the Virgin, and the miracle of the cupola's apparition is certainly connected with this religious tradition. The name of the Cathedral itself could not give a clearer indication of the feeling elicited by the suden vision of the cupola: it is the dome of Saint Mary of the Flower.

Modern Domes

How much better than the masterbuilders of the past have modern engineers and architects been able to do? As far as domes are concerned, the tools of modern design, the electronic computer and model analysis, have allowed today's average technologist to achieve results that few geniuses of the past could have dreamed of. This does not imply that modern domes are more significant or more beautiful than those of the past (most are not), but only that their dimensions far exceed those of the domes built before the industrial revolution. Actually, one could easily prove that while technology has leaped ahead in devising new ways and materials for the erection of monumental domes, architecture has not kept up with it and that, except for the works of men like Nervi and Saarinen, no modern roof can compete with Hagia Sophia, Santa Maria del Fiore, or Saint Peter's.

The largest span roof built to date is the triangular-base, concrete dome of the Centre National des Industries et des Techniques in Paris designed by Zehrfuss and engineered by Esquillan in 1968 (Fig. 13.12). It consists of three enormous buttresses, springing from the corners of an equilateral triangle with 720-foot sides and meeting 152 feet above the center of the triangle. The three buttresses constitute a double dome with a lower shell and an upper shell each only two-and-one-half inches thick, connected by vertical diaphragms of the same thickness set in a rectangular pattern (Fig. 13.13). The upper shell is corrugated across the direction of the buttresses to increase their buckling-strength. The three corners of the dome were poured on wooden scaffolds. The upper parts of the buttresses were built without centering. Movable steel forms were used for the inner shell while the upper shell was poured on wood fiberboard forms supported by the vertical diaphragms. The central portions of the three buttresses were poured simultaneously, starting at the three corners, and became self-supporting upon meeting at the top of the roof. The lateral portions of the three buttresses were then poured, widening the buttresses to their final dimensions in three separate operations (Fig. 13.14).

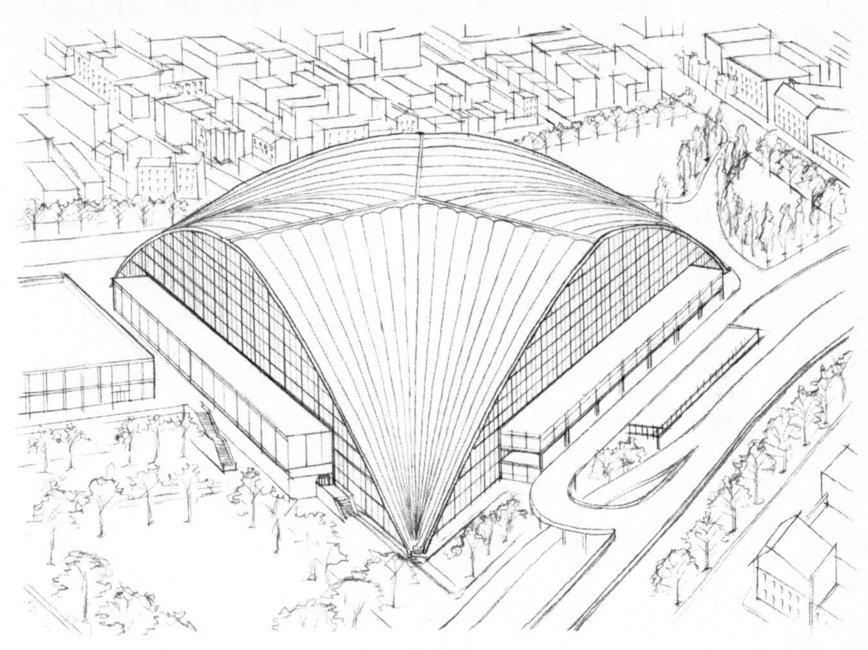

13.12 THE C.N.I.T DOME IN PARIS

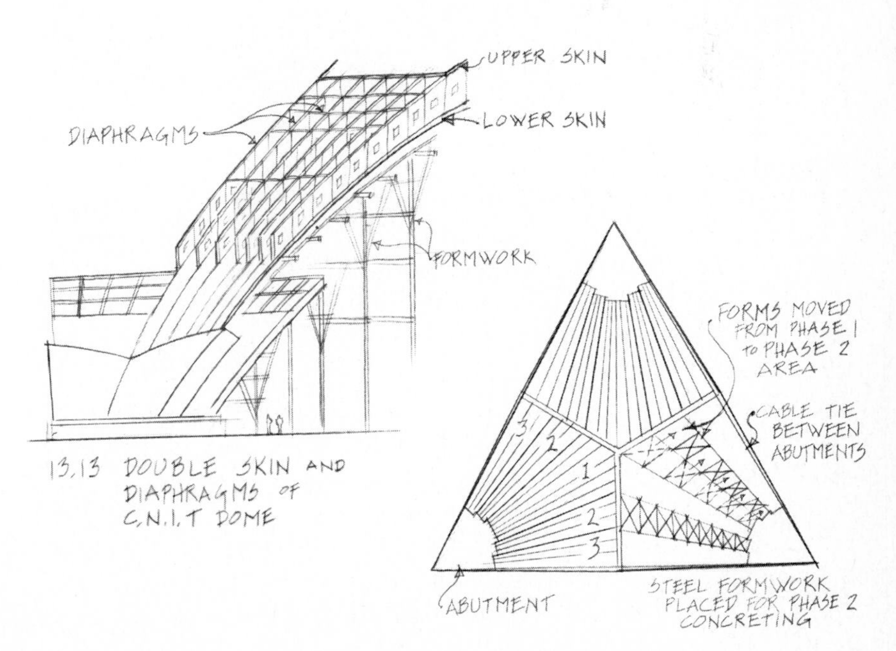

13.13 DOUBLE SKIN AND DIAPHRAGMS OF C.N.I.T DOME

13.14 POURING SCHEDULE OF C.N.I.T. DOME

The three-groined vault thus obtained exerts large outwards thrusts at its corner supports. These are absorbed by cable-ties connecting the three corners and post-tensioned in installments as the roof was erected and its weight and thrusts increased. It is hard to describe the number of innovative ideas that went into the construction of this technological masterpiece or to give a feeling for the monumentality of this structure. Suffice it to say that it covers over five acres and has one million square feet of exhibit area.

Why should it be that the largest dome in the world is built of concrete rather than steel? The reason for this apparent contradiction is that the dimensions, shape, and materials of our monuments are always dictated by the tyranny of economy. Depending on the availability of materials and of specialized manpower, on fabrication procedures and engineering traditions, concrete may be competitive in a given country or location at a given time, while steel may be more economical at another location or time. If the architect is sometimes inclined to put beauty first, the engineer never forgets that economy must prevail for a project to become reality. The fact is that our monuments are more often engineering rather than architectural achievements, except when the designer is at

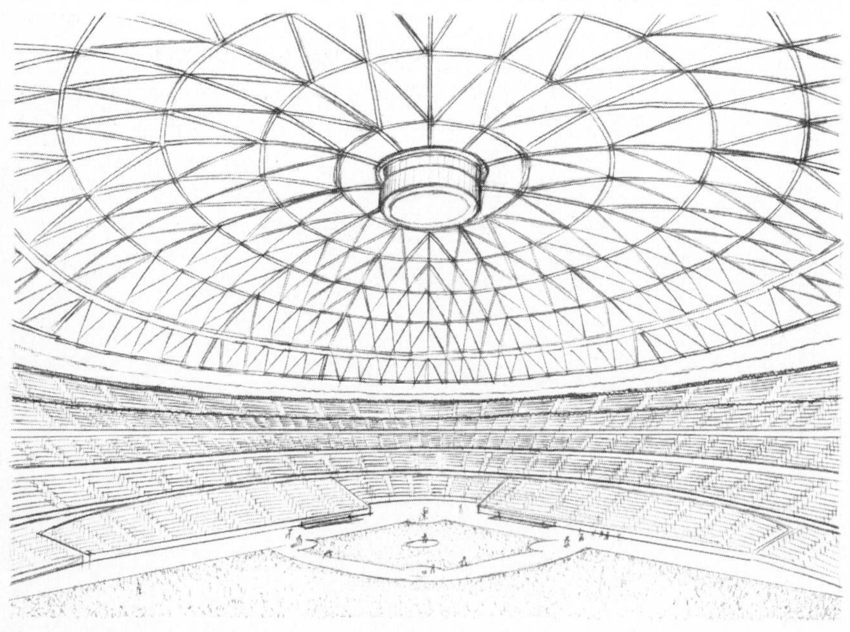

13.15 THE ASTRODOME IN HOUSTON, TEXAS

the same time a great engineer and a superb architect. (Perhaps the only man to fulfill these contradictory requirements in our time was Pier Luigi Nervi.) Thus the largest dome built to date in steel is the Louisiana Superdome, which covers a stadium rather than a church, and the second largest, the Astrodome (Fig. 13.15), is also a roof under which any kind of sports can be played. It would be naive and dishonest to deny the technological feat of erecting one of these roofs, covering nine to ten acres and up to 70,000 spectators, and employing materials of such strength that their weight is less than thirty pounds per square foot of roof. (Brunelleschi's inner dome alone weighed 700 pounds per square foot.) Neither can one deny the technical elegance of their structural design, the ingenuity of their construction procedures, or the immense feeling of space they create. Even so, though their dimensions call forth admiration, we are not moved by the space they create. When the dimensions of a structure are so large that they cannot be grasped through an act of pure intuition, our capacity for emotion is stunted. We may still be awed by a mountain or by the waves of a tempestuous ocean. These are the works of nature. But we seem to be unable to grasp the greatness of our own achievements unless we participate in their realization or they are interpreted for us in the language of the artist or the philosopher.

The triumphs of modern technology become meaningful to us only when they can be admired from such a distance that their dimensions become human. The beauty of the Verrazano Narrows Bridge, of the New York skyline, or of the C.N.I.T. exhibition hall can only be perceived from a distance. And yet one knows that we shall strive towards greater heights, larger spans, and wider areas, driven by many impulses the most important of which is and will always be of a spiritual nature. Each era has expressed this impulse through different means, but nobody can doubt that the spirit of an age is clearly expressed by its architecture. Since there cannot be architecture without structure, let us praise the hidden skeleton that allows us to realize beauty through the collaboration of architecture and technology.

14 Hagia Sophia

The Construction of the Church

We owe the miracle of Hagia Sophia to the vision of Justinian the Lawgiver, fourth Christian Roman Emperor, and to the ruthlessness of his wife, Theodora, a former circus performer and international courtesan. But the erection of this most famous of all Byzantine churches (Fig. 14.1), the only Christian church in the world used uninterruptedly through fourteen centuries to worship God the One, came about because of an unexpected and tragic event.

For over thirty years Justinian had dedicated himself to two goals: to stem the pressure of the barbarian invasions at the periphery of his wide empire and to create at its eastern heart a lasting monument to himself. By the time of his death in A.D. 565, at the age of eight-two, he had been more successful with the second than with the first: the barbarian tide had been barely stemmed, but over a thousand monuments and churches had been built in the city founded by Constantine the Great in A.D. 330. The achievement of Justinian's two ambitious goals demanded large amounts of money, and taxation under Justinian had been high and unevenly spread. The populace, for these many years docile under this financial burden, revolted in 532. With the cry of "Nika" (conquer) they took over the city, looting and burning everywhere. Justinian, an introverted and cultured man, ready to compromise, spoke to the people massed in the Hyppodrome, but to no avail. He might have fled the capital at this time, had it not been for the stand taken by Theodora, who swore

14.1 OUTSIDE VIEW OF HAGIA SOPHIA

that she would not flee even if the Emperor abandoned his throne. Reassured by her determination, Justinian decided to put down the rebellion. When the fighting ended 30,000 men, women, and children lay dead in the streets, but not before the populace had set afire and destroyed the important church of Hagia Sophia, dedicated to the Holy Wisdom of Christ by Constantine II. Its wooden roof had made it easy prey to the flames.

It took Justinian just thirty-nine days to decide that the reconstruction of this church was to be done on a scale unprecedented in human history and to set forth on the path. The man entrusted with the realization of Justinian's dream was Anthemius of Tralles, a Greek from Asia Minor steeped in mathematics and physics, one of the greatest architect-engineers of all time. Anthemius's assistant was Isidorus the Elder, also a Greek, from Miletus.

On February 23, 532, the erection of the church began with 100 overseers and a force of 10,000 workers. The workers were divided into two gangs, one for the northern and the other for the southern half of the building, so as to create competition and simultaneous progress. At the end of the day the workers were allowed to search for coins buried in the excavated soil as compensation beyond their regular wages. One feels that there is in all this more than meets the eye, for the construction of the church had obvious social overtones concerning unemployment and wages. Justinian had surely requested his architect for plans of the new church well before the destruction of the old, homonymous one. Even the genius of Anthemius could not have conceived and produced them in such a short time.

Justinian roamed the site day and night dressed in work clothes, giving advice and encouragement. At a time when the uprights of the scaffold for one of the great arches seemed in danger of splitting under the enormous weight of the arch and even Anthemius had lost his self-assurance, the Emperor advised him to continue the construction of the arch because "when it rests upon itself, it will no longer need the uprights under it."

Indeed, Hagia Sophia was Justinian's monument in all ways. He would go without food for a full day when engrossed in the work. He was able at times to sleep only an hour a night because, as Gibbon stated, "the body was awakened by the soul." It is no wonder that under his obsessive drive the construction of the largest Christian church ever built, covering 81,375 square feet and topped by the most daring dome ever conceived, was completed in five years, ten months, and four days. On December 27, 537, Justinian, after riding from his palace on a white horse, dismounted in front of the church, took the hand of the Patriarch of Constantinople, and walked with him through the atrium and the narthex, entering the magnificently decorated, immense nave. At this point he dropped the Patriarch's hand, ran to the center of the church, and looking at the floating dome, cried, "Glory to God, who has deemed me worth of fulfilling such a work. O Solomon, I have surpassed thee."

Thus was inaugurated fourteen centuries ago the Justinian version of Hagia Sophia, a church revered by people of all races and religions, sung about in Latin hexameters by the poet Paul the Silentiary, written about by visitors from all corners of the world, and, paradoxically, thoroughly known only in recent times, after it stopped functioning as a house of worship.

The Interior of the Church

There are many reasons for this universal admiration. The interior of Hagia Sophia during its nine centuries of Christian Orthodox worship was a space of almost incredible opulence (Fig. 14.2). Marbles gathered from all parts of the Empire paved the church floor, the pieces dovetailing

14.2 THE INTERIOR OF HAGIA SOPHIA

so as to create a flow of color that reminded visitors of the sea. Four monolithic green columns grew from the floor to support each of the lateral walls under the great north and south arches. Above these rose six red columns at the level of the women's galleries, and the entire walls were covered with marble below and mosaic above. Areas outside the main nave were populated by columns, 250 of them, of the finest marbles. They supported barrel vaults, also covered with mosaics. At the entrance to the church (Fig. 14.3) the *narthex* vaults, all four of them, were covered with mosaics, the large center one representing Christ the Infant on the lap of his Virgin Mother. The narthex was followed by a large space, roofed by a half-dome supported on marble columns, and the apse at the opposite end of the church was closed by another half-dome pierced by five windows and encrusted with mosaics. In contrast to the dark side galleries, supported by smaller decorated vaults on columns, the center space of the church, a one-hundred-foot square expanse with enormously massive pillars at its four corners, was lit by the forty windows at the base of the dome. The light from these windows contributed to the sensation that the shallow spherical dome floated over the church rather than being supported by the four great arches and four curved surfaces, the pendentives, that led from the square plan defined by the corner pillars to its own circular base. As Procopius wrote, "it seems not to rest on solid masonry, but to cover the space with its golden dome suspended from Heaven." The four pendentives were covered with mosaics representing four angels, while the entire dome was a golden surface with a large figure of Christ the Creator of the World—Pantocrator—floating over the center of this immense space. When the Patriarch and the Emperor entered this central, warmly illuminated area, they must have looked superhuman to the populace crowding the darkside galleries and floor spaces. But the people themselves were to enter this miraculous area shortly thereafter to accept the sacrament of communion and to become equals to the heads of Church and Empire. The new state religion and its rituals had both exalted the temporal power of the ruler and made him human in the eyes of the people.

The Orthodox rituals were no less magnificent at night. Daylight from the forty dome windows was then replaced by the light of eighty lamps attached to the base of the dome. In addition three circles of concentric lights hung from the dome. Oil lamps were supported on the lower columns as well as on the walls above the lateral arches and on the periphery of the church. All these lights, reflected by marbles and mosaics, created a truly magical atmosphere.

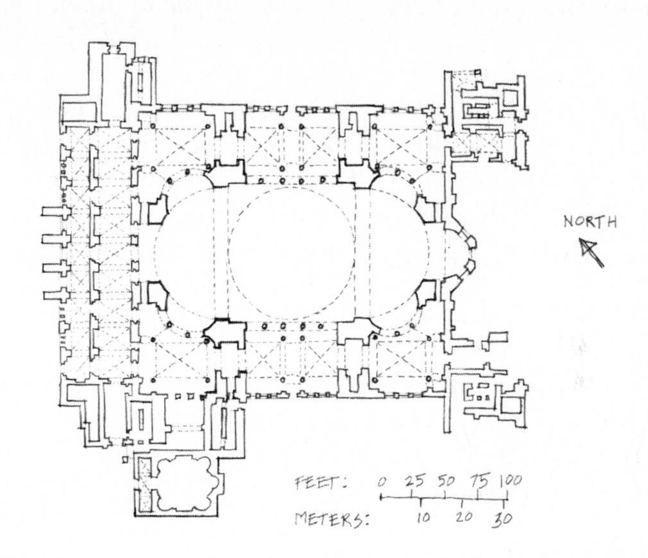

14.3 PLAN OF HAGIA SOPHIA

All architecture bears a message. The message created by Anthemius for his Emperor, his patriarch, and his people rang loud and clear. This space, a symbol of the protective love of the Church and Empire, is covered by curved surfaces, which embrace and protect the people humbly assembled to pray for the love of the great King of the Jews, God-made-man for their salvation. But the magnificent interior does also signify the greatness of the state and gives assurance of its strength and magnanimity. The light supporting the dome made it into a "dome of heaven" and elevated the spirit to celestial thoughts, but also served to flood down the opulent walls and arches and domes, reminding worshippers of the richness of the Emperor's palace. Meanwhile the church orientation pointed to the rising sun and to the hopes of the world, and the altar under the eastern half-dome roofing the apse lay in its semi-darkness to increase the mystery of the mass. Seldom have two such contrasting messages as those broadcast by Hagia Sophia been incorporated in a single, harmonious, and mesmerizing architectural ambience. When such a goal is achieved, we are confronted with the greatest of architecture.

Anthemius, who bent all of his knowledge and vision to this achievement, at times allowed his artistic daring to overcome his technical judgment. The result was a series of structural failures he did not live to witness. Even so, it is the incredible contrast between the sensuous decoration of Hagia Sophia's interior and the almost abstract geometrical shapes of its structure that must be understood if one is to appreciate the greatness of Anthemius's conception.

The Structure of the Church

Unfortunately the successive accretions to the church's exterior, either required to strengthen it or added later for functional purposes, do not allow us to see the original construction of Anthemius. But we can reconstruct it.

Consider a square with four pillars at its corners (Fig. 14.4). The space between each couple of pillars is spanned by a full arch in the shape of a half-circle, whose rise equals half its span. A horizontal circle is supported on the crowns of these four arches and a dome rises from it. The original dome was not a full dome, with a rise of forty-five feet, but rose only to approximately twenty-five feet. Hence it was shallow, more like an upside-down dish. The dished dome was reinforced and made stiffer by forty equally spaced radial ribs meeting at the top of the dome, where they vanished. Forty windows were set at the dome's base, between the forty ribs. Spherical surfaces plugged the opening between the horizontal ring at the bottom of the dome and the halves of the arches at right angles to each other rising from each pillar. These spherical surfaces, the *pendentives,* were triangular in form, with their lower corner resting on the pillar. This was the essence of the structure supporting the original dome. Except for the dome itself it has remained unchanged to this day.

The lateral arches supporting the dome and open to north and south are partly filled by walls, which rest on four large green marble columns. These columns in turn hold up six smaller red marble columns, which support the walls. These walls were perforated by three rows of windows, and hence, together with the two rows of columns at their bases, presented a light,open apperance. The lower row of windows was plugged later, without substantially changing the walls' appearance.

The western arch on the entrance side and the eastern arch on the apse side are not plugged by walls but are flanked (outside the square plan) by two half-domes resting on the main two east and two west corner pillars and on two additional pairs of smaller pillars, to the east and to the west of the main east and west pillars (see Fig. 14.3). These large half-domes are in turn flanked by two smaller half-domes and a barrel vault. The barrel vaults are in line with the east-west axis of the church, and the eastern barrel vault is closed by a last half-dome, which constitutes the roof of the apse. All the smaller half-domes are supported, like the lateral walls, by two sets of columns set on semicircles.

Finally, outside the lateral north and south walls two aisles are created by longitudinal rows of columns connected by barrel vaults, and an outside wall encloses the entire church in a 200-foot square. The geometrical scheme of the structure is simple and seems to correspond to the scheme of the interior space. What remains to be explained is why the two lateral arches were plugged by walls, while the east and west arches were continued into a series of large and smaller half-domes, so as to give the church a longitudinal plan.

The reason for this lack of total centrality is to be found in the requirements of the liturgy as well as in the Byzantine tradition of church construction. Many prior Byzantine churches were built with a completely centralized plan and were roofed by wooden domes. (The first Hagia Sophia seems to have been built in conformity with this arrange-

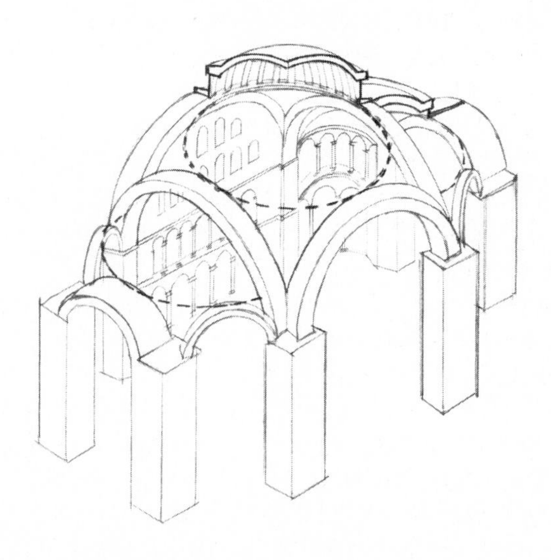

14.4 THE STRUCTURE OF HAGIA SOPHIA (AFTER R.J. MAINSTONE)

ment.) But the Roman structural scheme for buildings where large numbers of people gathered was that of the basilica, a longitudinal building with a wide central aisle terminating in an apse and lateral, usually narrower, aisles covered by a wooden truss roof. Anthemius, confronted with these two traditions as well as with the requirements of the mass liturgy, which called for the altar being situated at one end of the church, devised a compromise. The centered square structure was given an elongated appearance internally by screening the lateral arches with pierced walls and by prolonging the dome longitudinally by means of larger and smaller half-domes. In so doing he achieved an unencumbered span of 102 feet in the transverse direction and one of over 200 feet in the longitudinal direction, a span never before achieved. Thus was dimension made supreme and almost infinite to the naked eye. And to make the other human dimension—time—also infinite, he decided to avoid the use of wood in the roofing of the space. His Hagia Sophia was going to be fireproof. As a consequence of the avoidance of wood, the large spans could only be covered with arches, barrel vaults, and domes, shapes that allow the use of compressive materials like stone and brick. It is seen that, once Anthemius had decided to satisfy the needs of liturgy, to realize a grandiose space, and to aim at eternity, his choice of structure simply followed. And this correspondence between architectural requirements and structural needs is another characteristic of great architecture.

A good engineer, Anthemius used his materials wisely. He knew that the four main pillars and the four secondary pillars would have to support the weight of most of the structure and resist the thrust of the dome. Therefore he made them of granite, a heavy, strong stone. The horizontal surfaces of the granite blocks were carefully smoothed out to make sure the pressure on them would be evenly distributed. At critical joints sheets of lead were interposed between the blocks to insure perfect contact between them. Stone in the shape of columns was also used to support the walls, the half-domes, and the lateral vaults. But the vaults, half-domes, and the great dome itself had to be made as light as possible to reduce their weight and the consequent thrusts on the pillars and columns. Hence they were built of Byzantine bricks, measuring approximately a foot-and-a-half square and about two inches thick. The bricks were joined by lime mortar, a material which slowly becomes as hard as the brick and which smooths out any imperfections between adjoining brick surfaces. Enormous wooden scaffolds had to be built to support the arches, the vaults, and the domes during construction, until the mortar set and bound them into an almost monolithic structure.

In order to understand where Anthemius partly failed, one must become aware of both historical and structural causes. The Romans had perfected the arch, the vault, and the dome and had built domes as large as that of the Pantheon, 142 feet in diameter, by developing concrete, a material capable of standing a small amount of tension. In addition, most wisely, they made their curved roofs very thick and always supported domes all around their base. But the tradition of concrete dome construction had been lost. Moreover, for the first time in the history of the dome Anthemius conceived of supporting one (and the largest so far, at that) on four corner pillars. This was a revolutionary conception because of course domes, like arches, have a tendency to spread under loads. If they are not to fail by opening up at their base and cracking their supports must be restrained from moving outward. Hence the dome thrusts at Hagia Sophia had to be resisted by the four supporting arches and the thrusts of these arches by other structures in turn.

Anthemius knew all this, not in a strictly mathematical way as the modern engineer does, but more or less intuitively. He wisely buttressed the east and west arches by means of the large half-domes and these in turn by means of the smaller half-domes and vaults until the dome-thrust in the east and west direction was brought down to earth. If he could have used the same mechanism to resist the thrust in the north and south directions (as the Turks did later in the Blue Mosque), he would have encountered no problems. But the architectural needs of his space did not allow him this solution, and he had to invent makeshift remedies. He made the unpropped north and south arches smaller in span so they would thrust less, and thicker, so they would be stronger. In addition, he supported the north and south walls by a second hidden arch under the main arch. He finally propped up each main pillar by means of an additional buttress connected to the pillar by means of vaults (anticipating the system of buttresses used by the builders of the later Gothic cathedrals). The idea of filling the smaller north and south arches with lateral walls provides a measure of his greatness as an architect. Through an optical illusion, the walls give these smaller arches an appearance identical to those of the larger, open east and west arches and the difference in span is practically unnoticeable.

The Fate of Hagia Sophia

Anthemius knew he had designed and built the largest and most beautiful church in the world. He was aware of the tremendous structural problems Anthemius the engineer had overcome to allow Anthemius the architect to triumph. He died the recognized master-builder of all time. But only twenty-one years after the consecration of his masterpiece, following two earthquakes in 553 and 557, the eastern arch with the abutting half-dome and part of the main dome collapsed.

The intuition of Anthemius was insufficient to graps quantitatively the play of forces among the elements of the church, and yet the collapse might not have taken place were it not for the numerous earthquakes to which the church was subjected all through its life. The direct cause of the collapse was the insufficient buttressing of the main dome by the lateral arches: under the thrust of the dome and the shrinking of the slowly setting mortar, the north and south arches moved slightly outwards, allowing the supports of the eastern arch to spread, the crown to crack, and the half-dome to collapse.

It goes without saying that Justinian decided immediately to rebuild the dome. His new architect, Isidorus the Younger, a nephew of Anthemius's associate, aware of the enormous thrusts exterted by the shallow dome, rebuilt it in the shape of an almost full half-sphere, thus increasing its rise by twenty feet and reducing the thrust by thirty percent. The top of the dome now soared 180 feet over the level of the church floor. The church stands to this day with the profile given to it by Isidorus the Younger.

But the essential weakness of the original buttresses was still there. In 989 the western arch and half-dome collapsed. In an attempt to stem these failures, two enormous corner buttresses were built outside each of the north and south sides of the church under Emperor Andronicus the Elder (see Fig. 14.3). But these buttresses, which deface the exterior of the church to this day, were unable to prevent a second collapse of the eastern arch in 1346, following another severe earthquake.

By this time the knowledge of buttressing which had climaxed in the construction of the Gothic cathedrals in the West, had spread, and the repairs took only a few months. Finally in 1847 the church was strengthened by modern means under the supervision of the Swiss architects Gaspare and Giuseppe Fossati. They circled the base of the dome with an iron chain (see Chapter 13), and thus reduced the dome's thrust to the point where they dared to dismantle the upper parts of some of the more recent lateral buttresses. The church, made secure structurally, withstood other earthquakes without signs of weakness. The dream of Anthemius was finally realized, but ironically the Hagia Sophia that may last in aeternum is not his church. It is not even a church.

Through the centuries Hagia Sophia suffered more at the hands of men than from the forces of nature. First its interior was desecrated by the Christian sect of the Iconoclasts, who destroyed many of the mosaic images in order to "cleanse the church of corruption." Then the church was looted by the Christian crusaders from the West, who converted it for fifty-seven years to the Roman Catholic ritual. Finally, in 1453, Mohammed the Second, the youthful, great leader of the Ottoman Turks and the conqueror of Constantinople, converted Hagia Sophia into a mosque.

At first this conversion did not appreciably affect the interior of the church, since the cultured Sultan was so impressed by Hagia Sophia that he allowed it to remain untouched. While the Christian crusaders had looted the interior of its treasures, Mohammed, upon noticing one of his soldiers pry some precious marbles from its walls "for the glory of the faith," smote him with his sword and thus made his reverence for the church clear to all his followers. He even left its name unchanged: Hagia Sophia became the Aya Sofia Mosque. Its appearance was changed only on the exterior and was possibly improved by the four minarets successively built by three later Sultans.

But when in the middle of the eighteenth century, in accordance with Koranic law, the mosaics were whitewashed, the building lost its opulent appearance and sent out a new message. With all images wiped out, with rugs covering the marbled floor, with large round panels and Arabic inscriptions defacing the pendentives, with new lights hanging from the roof and reaching down over the heads of the worshippers, with the mihrab set in the apse off the church's axis to point towards Mecca, and with the muezzins' prayer echoing from the top of the high minarets, Hagia Sophia had become a severe and more spiritual monument, lacking the luster of its previous mundane glory and seemingly unconscious of the sultans who had transformed Constantinople into Istanbul. Paradoxically, the new naked interior enhanced the purity of the building's structure. Undistracted by the decoration now under whitewash, the visitor could take in the abstract lines of the dome and its supporting arches, half-domes, pillars, and columns better than one could ever have done before. This purity of lines, this unearthly quality, was part of the new message that for 500 years the building was going to send out.

Who would have thought that the miracle of Justinian was going to undergo one more drastic transformation at the hands of the Moslem conquerors; that the Fossati reconstruction was not to be final in our own time? And yet it came to pass that in 1932 the new progressivist leader of the Turkish Republic, Kemal Atatürk, in a sudden reversal of policy, closed the mosque and pinned on its main door a sign in his own hand announcing that "the museum was closed for repairs." Thus it was that under the wise direction of American specialists the whitewash could be patiently removed and some of the best known mosaics again saw the light of day. Some less known ones were also rediscovered following the directions of the Fossati brothers who, before whitewashing the interior again, had carefully copied and catalogued all the mosaics they had seen. It is a slow rebirth of the old interior, which unfortunately will never regain its original splendor since the damage from so many modifications cannot be entirely remedied. And yet some of its glory is being resurrected, and the ineffable Christian dream of Justinian and Anthemius is living again thanks to the modern leaders of a Turkish nation and to the art experts from the New World.

For the third time the church's message has been radically modified. Architecture is for people and can only be alive when people use it for the purposes for which it was built. The Hagia Sophia of the Christian mysteries, in which all the people participated, is gone. Gone is the monument to the Emperor's power, where the populace flocked to see the personification of their own glory. Gone the mosque symbolizing for five centuries the conquering culture of the Turks and their dark mysticism. The message of the building is now purely intellectual: those who are knowledgeable about its past may fantasize about it, and those who understand its pure structural essence can do so better now than ever before. It is only through the opening of the building to modern architectural and structural experts that some of its problems and their fascinating solutions have been understood. It is only because the building has been accurately surveyed for the first time that we know that the dome is slightly elliptical rather than perfectly circular; that the lateral walls are not plumb, but lean backward by a substantial amount; that the church axis is not rigorously oriented to the east. But even if some mosaics are resplendent again and a number of people of all religions and races visit the building, one cannot help but feel a lack of relationship between this incredible structure and the people entering this great void of a museum. Gone are the believers, the priests, the incense, the colors, the music, the choirs, the murmurs, the rituals. Gone are the great illusion, the dream and the hope.

But who knows whether the eternal Hagia Sophia of Anthemius, Justinian, and Mohammed will not live again?

15 Tents and Balloons

Tents

As this review of structures, their principles and the monuments they support nears an end we come full circle to the modern application of a system first used by the nomads of 8000 B.C.: the tent.

In the primitive tent, as in so many of our structures, the structural and functional components were clearly separated. The center pole, stayed by ropes anchored into the ground, constituted the structure, while the animal skins tied to the ropes performed the protective function against sun and rain. The great enemy of the tent was the wind. A strong wind could blow the tent away, and a modest wind could make it flap and vibrate, tearing it apart even if the structure was strong enough to resist the wind force. The basic remedy for this weakness of the fabric is to stretch it between the ropes and to put it in tension. A handkerchief held out of a fast moving car will vibrate or flutter in the car-induced wind, but its flutter is reduced if it is pulled tight with two hands. All the ingenuity used by man to increase the usefulness of the lightweight roofs we call tents has been aimed at pulling their thin fabric taut and has led in the last fifty years to triumphant results. The modern tent is used in some of the largest roofs ever built and promises to become a permanent feature of our landscape.

The use of the classical tent, hanging from stayed poles, continued through the centuries. Most of the decisions taken by the Greek heroes of Homer in front of embattled Troy or by Roman generals in Shakespeare's plays took place under tents of the most varied shapes and flamboyant colors. But not all tents of antiquity were supported by poles and limited in size. The August sun in Rome can be quite hot and the 50,000 Romans enjoying the bloody spectacle of the gladiators in the Coliseum were sheltered from it by a retractable canvas tent, supported by ropes spanning as much as 512 feet, the shortest distance between opposite points at the top of the amphitheatre's exterior wall.

A fabric, whether its material is natural, like cotton or hemp, or man-made, like nylon or vinyl, is a thin membrane of great flexibility. It has no bending stiffness and can only resist one type of stress: tension. When tensed or *prestressed* to prevent fluttering, it acquires a greater stability and can support a variety of loads without excessive deformations. A most useful example of a tensed membrane is that used to catch people jumping from high windows or roofs to escape fires. It consists of

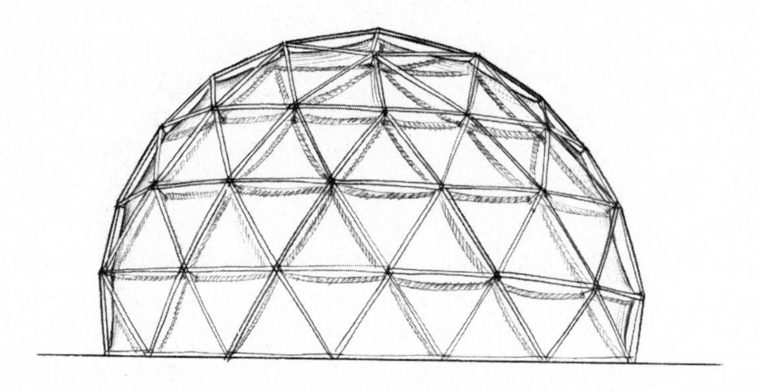

15.1a MEMBRANE TENSED INSIDE GEODESIC DOME

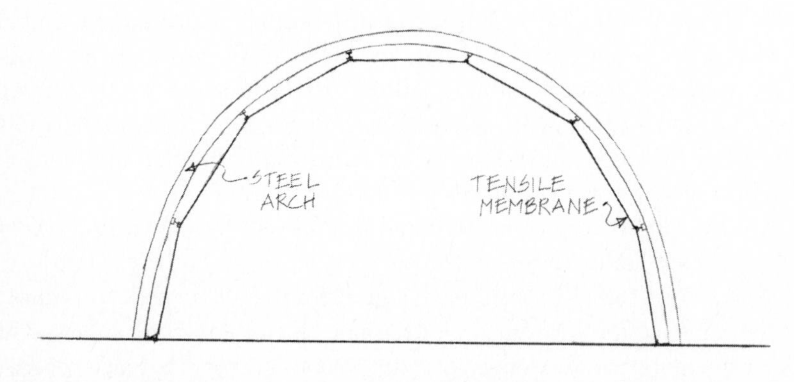

15.1b MEMBRANE TENSED FROM ARCHES

a ring of steel or aluminum inside which a circular canvas membrane is stretched by taut thin cables attaching it to the ring. People falling on these trampolins bounce up and down a number of times unhurt instead of impacting dangerously on a hard surface.

A flat roof could be built by means of a fabric trampolin, but it would deflect excessively under a load of snow or a high wind pressure. The next step in the development of a stiff canvas roof consists, therefore, in either stretching the fabric between the ropes of a tent, as is done to this day in the large tents used by circuses, or in hanging a downward curved membrane from an exterior space frame of bars and pulling it tight from the frame by means of ropes. Membranes have been tensed inside the bars of *geodesic domes* (Fig. 15.1a) (the lightweight frames invented by Buckminster Fuller that can be lifted by a helicopter) or from a series of parallel steel arches, to create warehouses, tennis court covers, and other fairly large enclosures (Fig. 15.1b).

Rather than pull the membrane from an outer frame, one may perhaps more efficiently pull down the membrane over an inner frame of steel. There is nothing new to this idea: it is the basis for the most commonly used membrane in the world, the umbrella (a Chinese invention introduced to England in the eighteenth century). The umbrella fabric is made taut by the curved-spoke ribs, which, in turn, are prevented from buckling, thin as they are, by the taut membrane that constrains their displacements. When membranes are spanned over steel arches, these can, similarly, be made much lighter and thinner than those of an outer framework. The Perma-Span system has perfected this concept for a variety of applications—from schools to tennis courts and from field hospitals to temporary offices on site (Fig. 15.2).

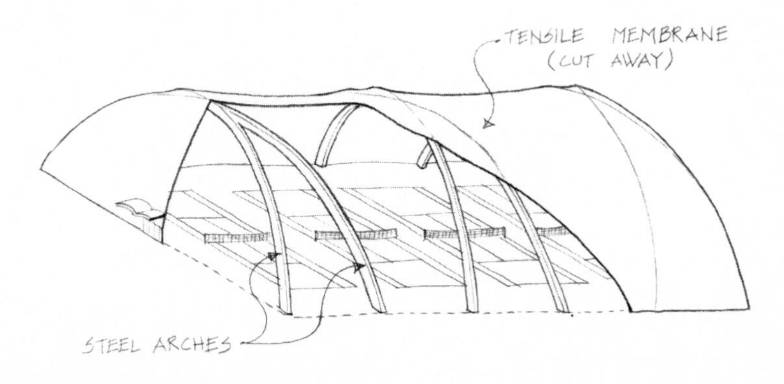

15.2 PERMA-SPAN MEMBRANE TENSED ON STEEL ARCHES

The tauter the membrane, the stiffer it becomes. Hence, fabrics made of natural fibers are not ideally suited to building large stretched tents. Their resistance to tension is limited while the larger the tent, the larger the prestressing tension needed to stiffen it. The extraordinary tents of the last fifty years have been made possible by the invention of plastic fabrics. These include the inexpensive vinyls that, unfortunately, deteriorate rapidly upon exposure to the ultraviolet rays of the sun, as well as the almost ideal glass-reinforced fabrics developed during the last decade. These consist of a network of glass fibers embedded in a thin layer of plastic. In its most recent type, the fabric is coated with Teflon (the same material used to coat frying pans) and, if required, by additional coats of reflective plastic materials. These composite fabrics reach very high strengths: a pull of more than 800 pounds for each inch of fabric is needed to tear them. Moreover, they are self-cleaning, since dirt does not stick to them, they do not catch fire up to temperatures of 1000°F., cannot be cut even with a hatchet, and can be given degrees of translucency all the way from clear transparency to total opacity. Their reflective properties make them good insulators against the heat of the sun in summer, while their translucency can be used in winter to allow the sun's rays to enter the space they enclose. Once again, chemical technologies of the most refined nature have come to help the structural engineer.

The use of the new miracle fabrics is limited by two factors. The first is their cost: they are about three to four times as expensive as the short-lived vinyls, although their useful life of twenty or more years compensates for this disadvantage. The second is their strength: high as this may be, it is not sufficient to span hundreds of feet, since the stiffness

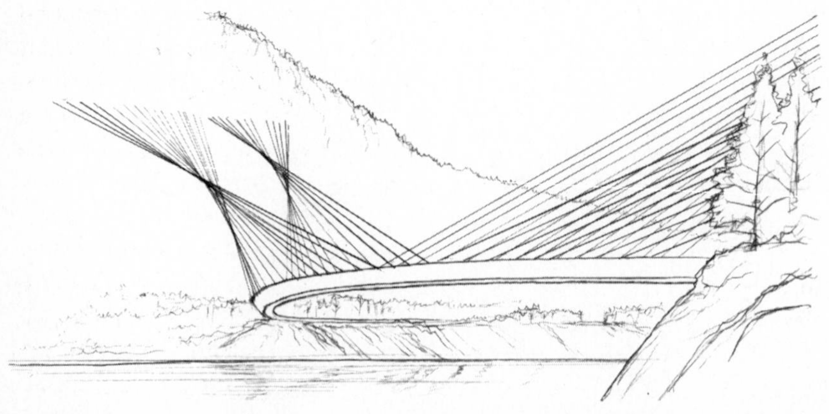

15.3 T.Y. LIN'S ROCK-A-CHUCKY SUSPENSION BRIDGE (PROPOSED)

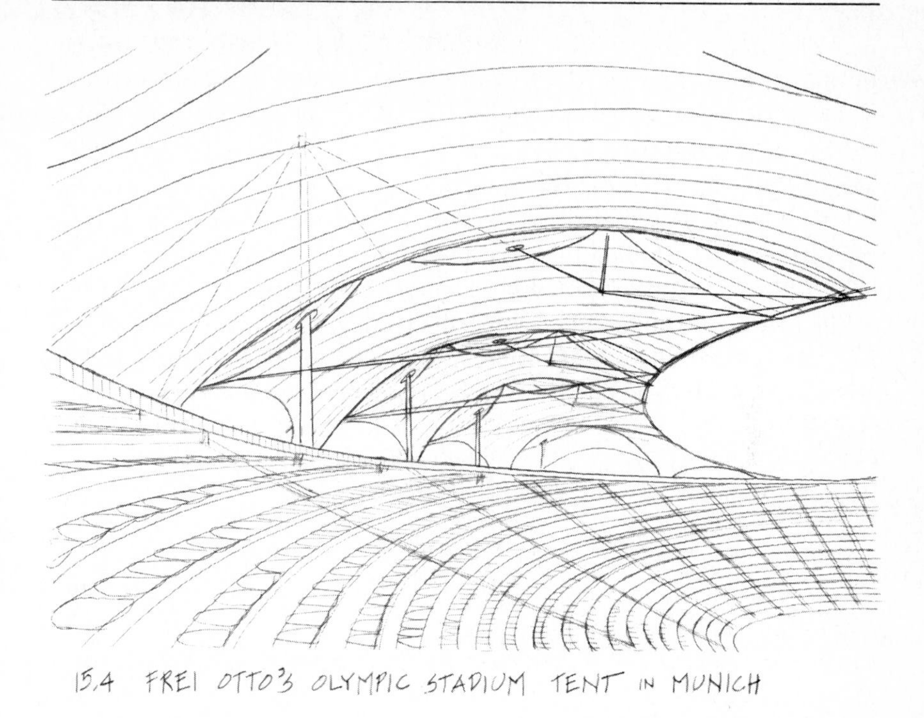

15.4 FREI OTTO'S OLYMPIC STADIUM TENT IN MUNICH

required by a membrane increases with its span. To overcome this last obstacle, man had but to look at nature. The spider, that most ingenious and patient of arachnids, has shown us how to spin a web of thin but strong threads, spanning distances extremely large in comparison to their diameter. The spiderwebs of man are made, of course, of steel cables, which are capable of supporting loads as heavy as those of a suspension bridge. Actually, the Ruck-a-Chucky curved suspension bridge, proposed by T.Y. Lin, the master of prestressed concrete in the United States, is supported by cables anchored at various points on the two banks of a river and looks as if it had been caught in a spider web (Fig. 15.3). Once a network of cables is suspended from suitable points of support, the miracle fabrics can be hung from it and stretched across the relatively small distance between the cables of the network. The German architect Frei Otto has pioneered this type of roof, in which a net of thin cables hangs from heavy boundary cables supported by long steel or aluminum poles. Following the erection of the tent for the West German pavilion at Expo '67 in Montreal, he succeeded in covering the stands of the Munich Olympic Stadium (Fig. 15.4) in 1972 with a tent that shelters eighteen acres, supported by nine compressive masts as high as 260 feet and by boundary prestressing cables of up to 5,000 tons capacity. (The spider, by the way, is not easy to imitate—this roof required 40,000 hours of engineering calculations and drawings.) The largest area covered by tents to-date (1982) is in Mecca. Horst Berger's project for sheltering the Moslem pilgrims occupies one million square feet or twenty-three acres.

One lovely, and necessary, characteristic of membrane roofs is their geometric shape. Holding a handkerchief by two opposite corners and trying to prestress it by pulling on the other two, it is easy to realize that the handkerchief can only be stretched by pulling up on two corners and down on the other two. The shape acquired by the handkerchief is that of a saddle, curving up between the corners held up and curving down between the corners pulled down (Fig. 15.5). Luckily such *saddle surfaces,* the only surfaces capable of stretching a membrane, are inherently pleasing to the eye.

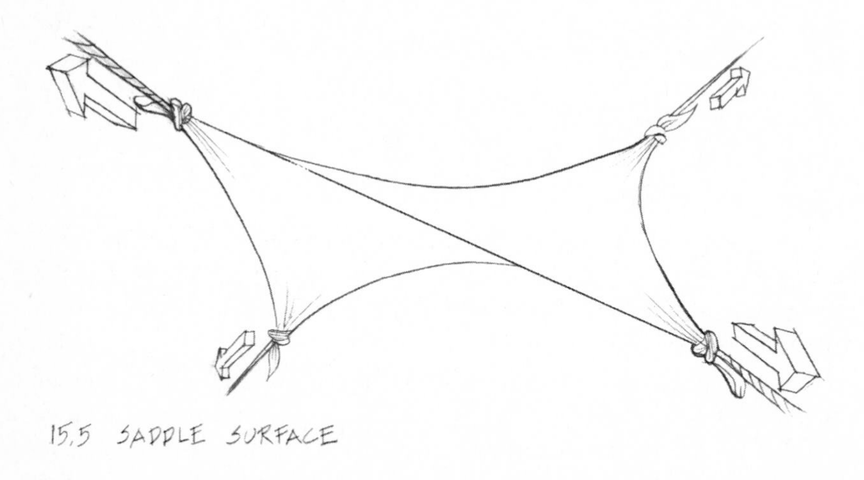

15.5 SADDLE SURFACE

In the continuous and triumphant development of larger and larger roofs supported by tensed elements, the time came when the network of cables became the primary structural component and the membrane was replaced by other materials. In 1950 an amazingly clever scheme for supporting a cable network was suggested by the Polish architect Nowicki for the "Cow Palace" in Raleigh, North Carolina. Its principle, simplicity itself, is an extension of that used by the common "director's chair," a canvas seat, supported by four wooden legs, crossed and pivoted at their midpoint, two in front and two in back of the seat (Fig. 15.6a). The tension of the loaded canvas compresses the two sets of legs and this

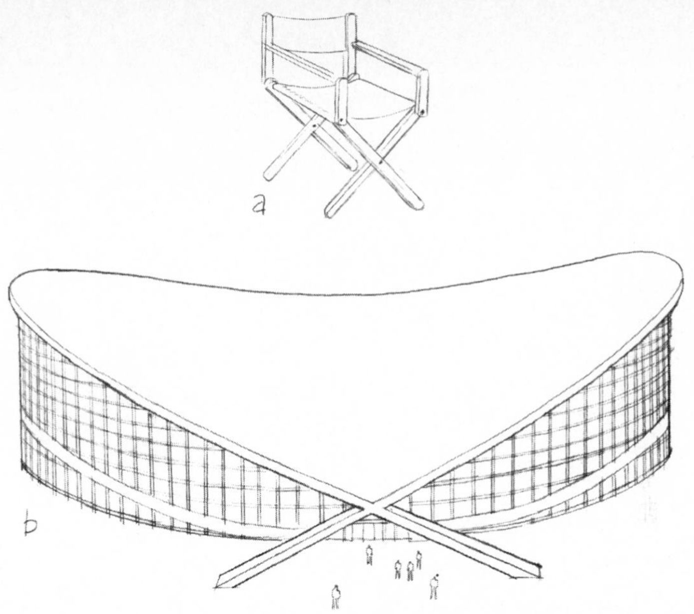

15.6a DIRECTOR'S CHAIR, 15.6b NOWICKI'S "COW PALACE" IN RALEIGH, N.C.

light, foldable contraption is capable of supporting the weight of a person. Nowicki replaced the two sets of scissored legs with two reinforced concrete arches, inclined at an angle of about 20° to the horizontal and pivoted at their two intersections (Fig. 15.6b). His network consists of two sets of cables: one curving up and parallel to the legs of the arches and another curving down at right angles to the first. He thus obtained a saddle surface of cables, which were set in tension, at least partially, by the weight of the arches hanging from them. The meshes between the cables were covered with undulating plates of translucent plastic.

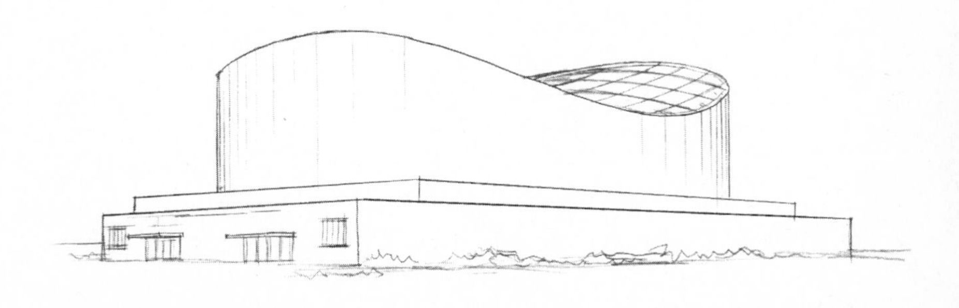

15.7 SADDLE SURFACE CABLE ROOF

15.8 DULLES AIRPORT CABLE ROOF BY EERO SAARINEN

This elegant and simple concept has been extended to a variety of suspension systems. Some consist of rings curved in space and supported by columns, carrying two sets of cables, one in the direction connecting high opposite points of the ring and the other connecting opposite low points, thus giving the network a saddle shape (Fig. 15.7). The roof itself consists of concrete slabs that tense the cables by their weight. The roof

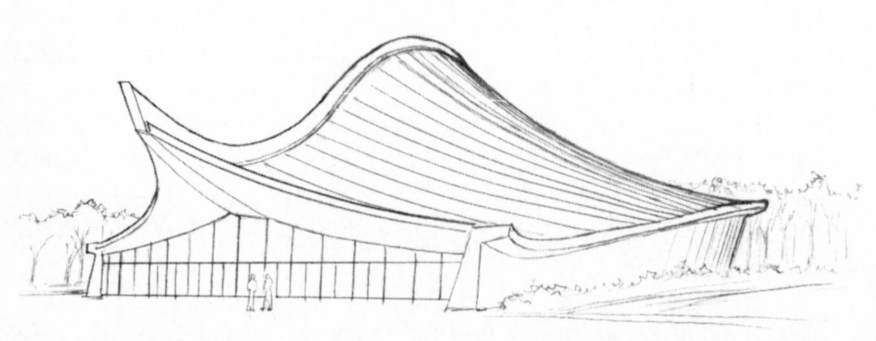

15.9 SKATING RINK AT YALE UNIVERSITY BY EERO SAARINEN

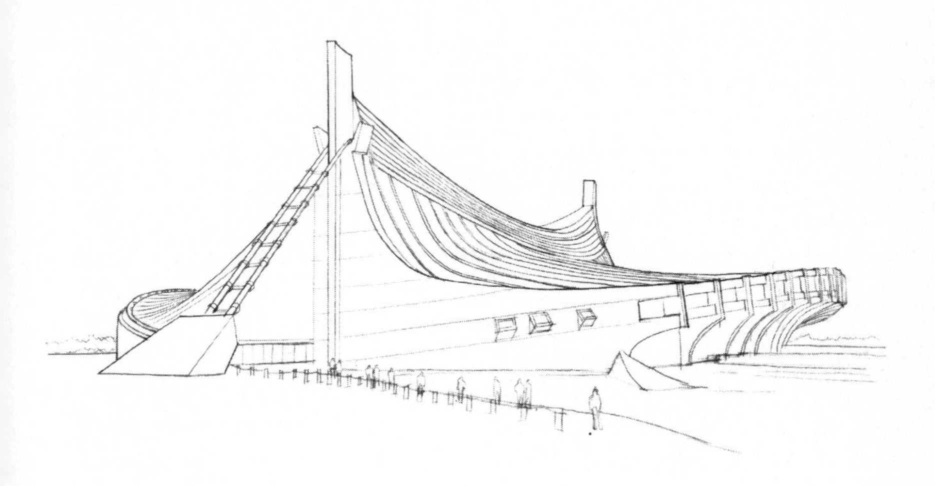

15.10 SPORTS PALACE IN TOKYO BY KENZO TANGE

of the Dulles Airport terminal in Washington, D.C. (Fig. 15.8) has cables running in only one direction, hanging from the top of opposite inclined buttresses and prestressed by the weight of the concrete roof. It is essentially a series of suspension bridges, very much like the Borgo paper plant of Nervi and Covre (see Fig. 9.20). In the skating rink at Yale (Fig. 15.9), designed by Eero Saarinen, the cables hang from a central concrete arch, are anchored to the curved walls of the rink, and are stabilized by the weight of the roof. Perhaps the most stunningly beautiful tent roofs of this kind are the roofs designed by Kenzo Tange for stadiums of the 1964 Olympics in Tokyo (Fig. 15.10),* in one of which the cables are supported at one end by a sculptured, off-center tower and anchored at the other to the walls of the stadiums.

Have we reached the end of the line with tensile roofs? No. The unquenchable desire for bigger and cheaper roofs and the inventiveness of the very best engineers have already led to a quantum leap in the conception and construction of tensile roofs. And the last structural material to enter the field is, of all things, air.

* The last significant improvement to the cable roof, achieved by the Uruguayan engineer Leonel Viera, is discussed in Chapter 16.

Pneumatic Structures

There is nothing new in the use of air as a structural material. We all ride in cars or on bicycles using inflated tires, which acquire their stiffness through compressed air. We play with tennis balls or footballs, both light and stiff because of air pressure. On the other hand, only during the last few decades has air become an important component in the structure of, first, small temporary buildings and, more recently, of permanent large buildings.

Actually air pressure can be used not only to inflate a closed balloon, but to prestress tents, by means of negative pressure or *suction.* Weidlinger was the first to apply this principle, in the complex of cable-strengthened tents at the Boston Zoo (Fig. 15.11). These hang from tripods of parabolic arches and are anchored to their peripheral walls. Fans pushing air *out* of the tents cause a depression that sucks in the tent's fabric. Thus, the tents are prestressed by the pressure of the outside air with respect to the minor vacuum created inside the tents.

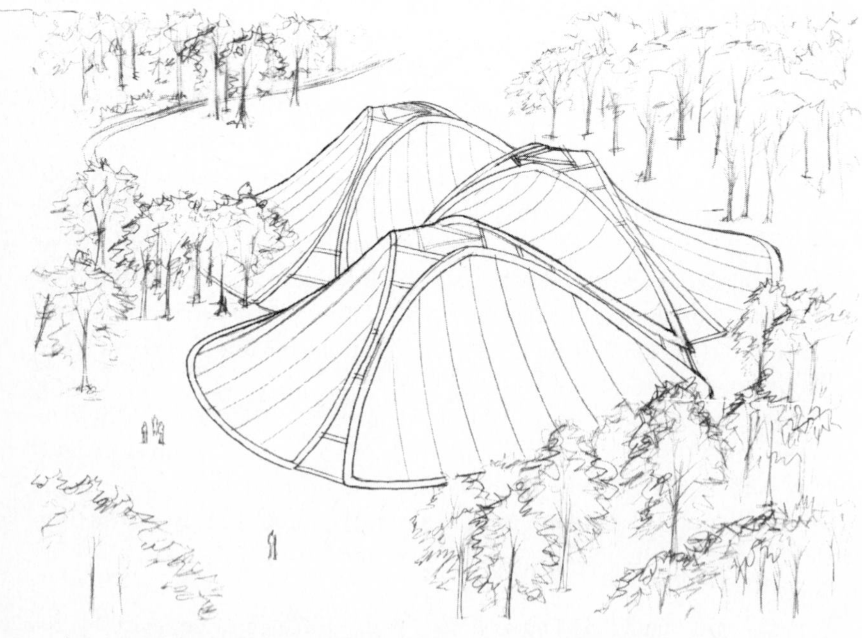

15.11 THE "SUCKED-IN" TENTS OF THE BOSTON ZOO

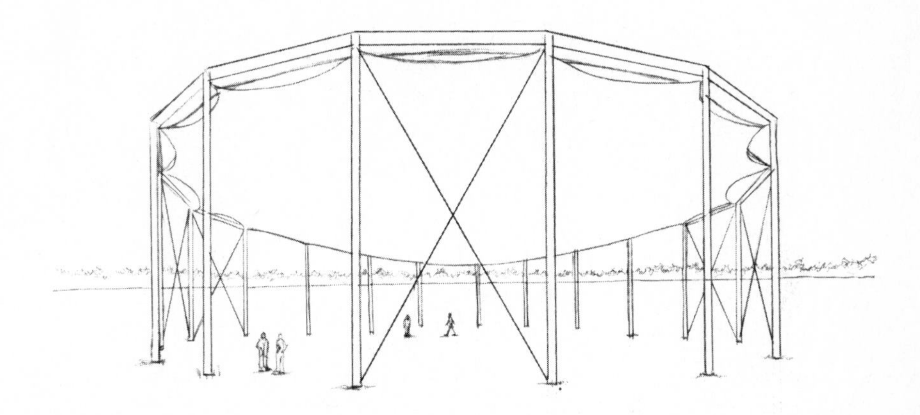

15.12 THE BALLOON ROOF OF THE BOSTON ART CENTER

The first proposal of the use of an inflated balloon as the structure for a large building came from the English engineer F.W. Lanchester in 1918. He patented a system "for an improved construction of tent field hospitals, depots and like purposes" in the shape of spheres or cylinders and actually designed a spherical balloon roof with a diameter of almost 2,000 feet, which he did not live to see built. He planned to stiffen the membrane with outside ropes and described the air-locks needed for entering and exiting the balloon. His drawings look very much like those of a modern air-supported tennis cover. But artists often foresee the technical ideas of engineers and H.G. Wells in 1895 described how the hero of his novel, *When the Sleeper Wakes,* after a long sleep, finds that cities are covered with transparent balloons reinforced by steel cables. In a sense, engineering could be defined as the realization of Utopia.

The first *pneumatic roofs,* so-called from the Greek word *pneuma* meaning breath, used spaces totally enclosed by membranes and made rigid by compressed air. The roof of the Boston Art Center (Fig. 15.12), designed in 1959 by the architect Carl Koch and engineered by Weidlinger, was built as a rapidly demountable, inflated roof, consisting of two circular membranes of nylon, 145 feet in diameter, zippered together and cranked up to and down from a ring of steel supported on steel columns. It

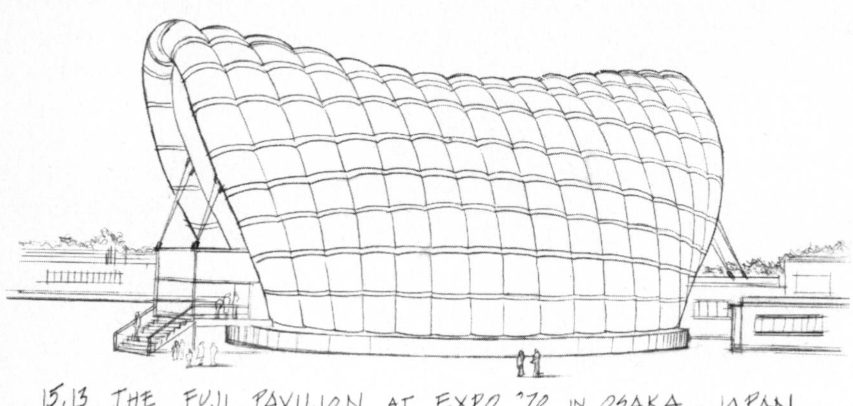

15.13 THE FUJI PAVILION AT EXPO '70 IN OSAKA, JAPAN

covered 16,000 square feet and 2,000 spectators. This inflated lens-shaped nylon balloon was proposed also as formwork on which to pour eventually a permanent reinforced concrete dome, an application of a principle which became successful years later (see Chapter 11).

And who would have thought that roofs could be built with the sausage shape of children's Christmas balloons? But this is exactly what Yutaka Murata did at Expo '70 in Osaka, Japan, where the Fuji Pavilion (Fig. 15.13) was built by means of 16 bent sausages, 12 feet in diameter, set around a 150 foot circle and reaching a height of 75 feet. The membrane for the tubes was not yet a modern fabric, but consisted of a plastic (polyvinyl acetate) of fairly high strength, coated with Hypalon paint on the outside and covered with a layer of another vinyl (polyvinyl chloride or PVC) inside. By increasing the pressure inside the tubes, the roof could be made stiff enough to resist winds of typhoon strength.

The concept of pressure-stiffened tubes has been theoretically extended by the Australian architect J.G. Pohl to a cylindrical vertical tube of transparent plastic, pressurized to make it stiff and capable of standing up as a column. The floors of the building, hanging from the top of the tube by means of steel cables, would be supported by the internal air pressure, which would have to be as high as one pound per square inch for each floor to be supported (Fig. 15.14). Numerous tests have been conducted to prove the feasibility of such a structure, but no realization seems to be in sight.

Smaller inflated mattresses have been used for a variety of purposes: all the way from those we sun ourselves on while floating on water to the mezzanine balcony of a theatre on which the spectators can sit and bounce up-and-down if and when the show becomes exciting. Inflatable rafts, with sides of tubing and a mattress floor, are the standard life-saving boats for ocean-crossing planes; one was used to cross the Atlantic under sail by a solitary French navigator. Probably, everybody remembers the unlucky lighter-than-air Zeppelin, prestressed by both compressed hydrogen and an inner aluminum frame, but few are aware that an inflatable small plane, with both fuselage and wings stiffened by compressed air, has been flown. Deflated, it is small enough to be stored in the trunk of a car.

Mr. Lanchester's design for a field hospital in 1918 probably could not have become a reality in his time for lack of strong fabrics, but since then the dimensions of pneumatic roofs have grown apace with the development of plastic fabrics. The first impulse for the use of huge balloons came from the military: they needed to cover the newly invented radars—plane-detecting electronic antennas in the shape of dishes with diameters of 200 or more feet—in order to avoid wind-produced deformation of their surfaces. The pioneering efforts of Birdair Structures met these needs by means of spherical balloons, strengthened by a triangular

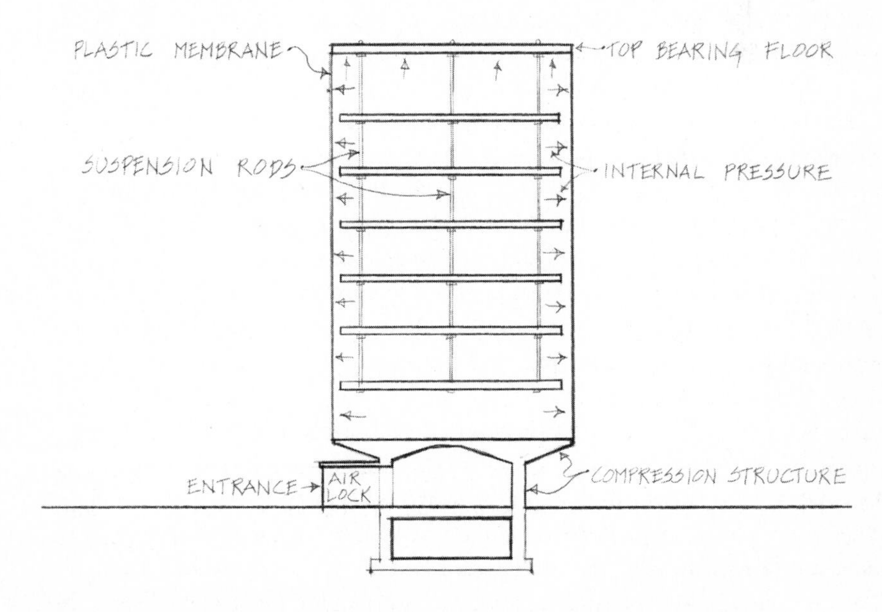

15.14 PRESSURIZED CYLINDRICAL BUILDING SCHEME

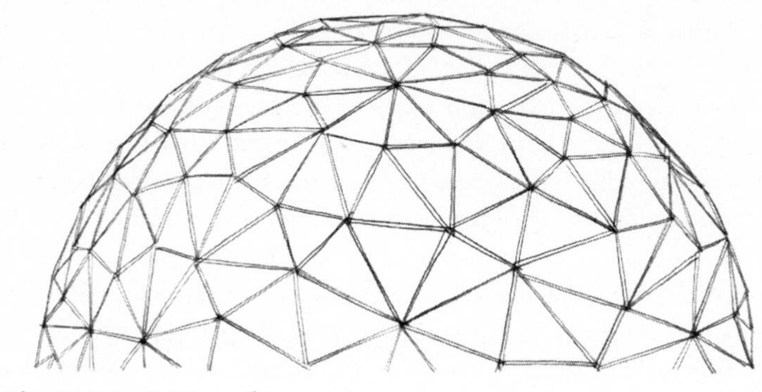

15.15 PRESSURIZED RADOME

net of cables that was superior to any other type of net in maintaining the shape of the pressurized balloons. Air-supported "radomes" (Fig. 15.15) now dot the United States and other countries. The efforts of the fabric manufacturers kept pace with the market for their strong membranes and soon glass-reinforced fabrics embedded in plastic "matrices" were produced by Owens/Corning. They opened up a new world to the requirements of architecture.

One man is the pioneer of this development. David Geiger was challenged in 1967 by the architects Davis and Brody to conceive a roof covering 100,000 square feet for the U.S. Pavilion at Expo '70 in Osaka within the limitations of a budget of 2.6 million dollars. He proposed a Lanchester-type pneumatic roof with clear spans of 262 feet by 460 feet, made of a Fiber-glas reinforced vinyl membrane and anchored to an inclined concrete wall or *berm* of ovoidal shape. It was supported by an air pressure of only three hundredths of a pound per each square inch, but could resist winds of 150 miles per hour. It became the wonder of Expo '70 (Fig. 15.16).

Geiger's conception was original in all the components of the pneumatic roof. The fabric was of a new type, the shape of the covered area and hence the shape of the ring to which the membrane was attached, was new. The lozenge type of cable mesh and its hardware were new. The use of a berm was new. The U.S. Pavilion was a quantum leap in the type of pneumatic structure, which has developed phenomenally in the years since Osaka.

How does a pneumatic roof work? Its basic components are the membrane with its stiffening cables, the ring they are attached to, and the fans that blow the compressed air in. As the membrane is inflated, the cables, which are usually suspended from it, become tensed and pull on their anchorages in the ring. The pull of the cables sets the ring in compression, (as the tensed spokes of a bicycle wheel compress the rim), so that it is usually built of inexpensive reinforced concrete. The air pressure supports the weight of the membrane (less than five pounds per square foot) and stiffens it against the action of lateral winds. Contrary to what happens in almost all other structures, the wind flow on the shallow, curved shape of the roof does not create a pressure on the top of it, but, on the contrary, an upward suction over almost its entire surface, just as the wind does at the back of a rectangular building. The largest pull to be resisted by the cables is due more to this upward suction of the wind than to the air pressure inside the membrane. The outer concrete ring serves not only as anchorage for the cables, but prevents with its own weight the lifting of the entire roof by the inner air pressure and the outer wind suction.

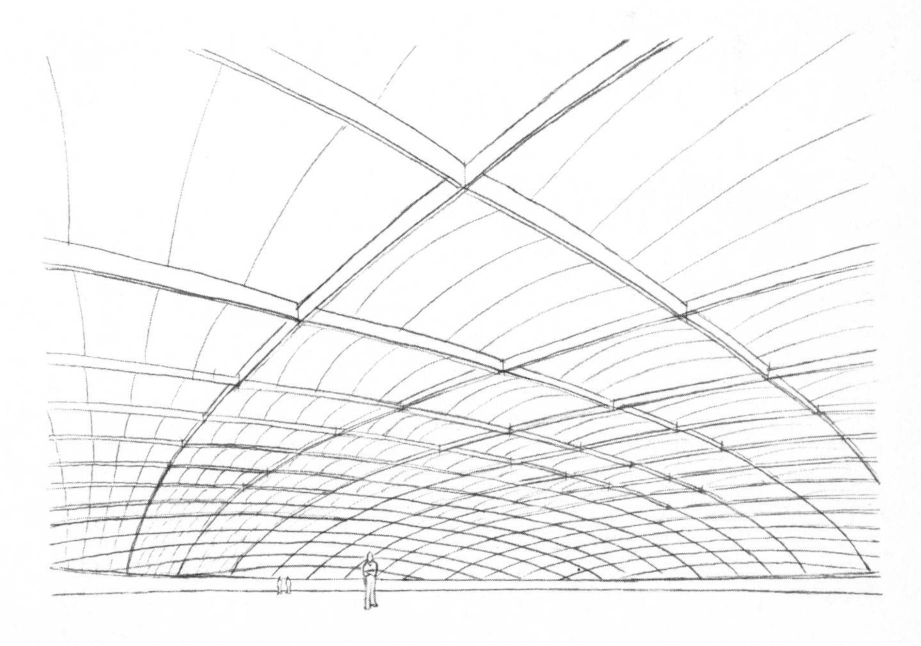

15.16 THE U.S. PAVILION AT EXPO '70 IN OSAKA, JAPAN

The largest roof built so far (1980) with a cable-reinforced, Fiber-glas Teflon membrane is the cover for the Pontiac Stadium in Pontiac, Michigan (Fig. 15.17). Designed as an afterthought to reduce the cost of the building, the roof covers ten acres and 80,400 spectators. It is attached to an outer polygonal compression ring high above the field and will hang from it, even if the roof develops a hole and collapses. (The hole would have to be 700 square feet in area before this could happen, since the fans can increase the airflow in case of an accident.) Its initial inflation took two hours and its cables were ferried across the stadium from one anchorage to the opposite anchorage by helicopter. Its inner pressure is only 3.5 pounds per square foot, which is one fourth of the maximum wind suction on the roof. Its use reduced the cost of the roof by sixty-six percent. Finally—another original idea of Geiger—its membrane consists of separate panels zippered to one another. Thus, when a tornado blew out a section of the stadium walls, suddenly decreasing the pressure of the inside air, the collapsed membrane flapped, tearing a few panels, but the torn panels were taken down, new reserve panels were zippered in and the stadium was ready within three days for the next football game.

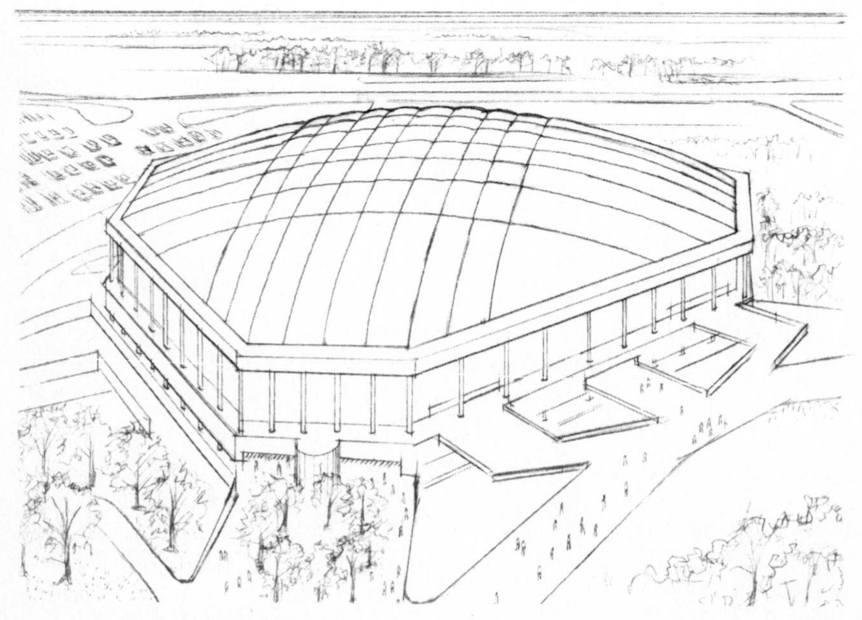

15.17 THE PONTIAC STADIUM IN PONTIAC, MICHIGAN

In our age of contemplated energy scarcity, the pneumatic roof has demonstrated interesting energy-saving properties. In the extremely hot climate of Riyadh, the capital of Saudi Arabia, the pneumatic roof of a recreational facility for the University could only be erected between sunset and sunrise as daytime temperatures of up to 140°F. made it impossible to handle the fabric. But once the roof was erected and before the air-conditioning system was activated, the temperature inside the facility was found to be thirty to forty degrees lower than the outside temperature, thanks to the reflecting properties of its white self-cleaning roof. On the other hand, in colder climates the availability of highly translucent fabrics, allows the southern sun's rays to penetrate the roof, heating the inside air and reducing demands on the heating system.

The major obstacles to expanded use of pneumatic roofs lie in the common inertia of the human mind. As in the case of tents, objections were raised that had to be answered before acceptance became wide. What if a hole is cut into the membrane by fire or by saboteurs? Modern membranes are fireproof and accepted for permanent structures by fire codes. Saboteurs would have to use electric saws with carborundum disks to be able to cut the fabrics. But even if this were to happen, an increase in the velocity of the blowers can always increase the supply of air and either keep the balloon up or allow it to come down so slowly that it may take hours or even days before the membrane collapses entirely. And moreover, if a flame so hot as to set the membrane on fire and burn a hole in it were to develop, the air under pressure exiting from the hole would snuff out the fire much as air puffed on a match. What about the inner air-pressure? Doesn't it make it uncomfortable to breathe? Nobody can notice the minute overpressure used to support the balloon, since it is smaller than the increase in pressure encountered in going down from the fortieth story of a building to street level. The spreading of the use of balloons all over the United States and the world seems to indicate that at long last most of the questions have been answered.

What can we expect in the field of pneumatic structures? Two recent studies may give an indication of the shape of things to come. The first was conducted to investigate the feasibility of covering with a pneumatic membrane an entire oil refinery, which was objectionable from a visual point of view among the lovely hills of Delaware. It had a diameter of 6,400 feet. The only problem encountered in its design was that of devising a drainage system for the water that would rain on its 731 acres of fabric.

The second study was a proposal for the federal government to cover with a single cylindrical pressurized membrane a complex of buildings four to six stories high, housing government agencies as well as other types of occupancies in a 400,000-square-foot area. The proposal showed that by constructing the buildings under the cover of the pre-erected membrane, construction is considerably speeded-up and cost decreased, and that a triple membrane can reduce the heating and air-conditioning requirements by an ingenious combination of membranes. Each panel of the outer and middle membranes is half-opaque and half-transparent. When the two membranes are separated by air pressure, the southern sun's rays can penerate at a slant through the transparent half-panels (Fig. 15.18a); when they are pushed together by the inner air pressure, the opaque halves of the middle and outer panels "close" the transparent half-panels and the roof reflects the sun rays (Fig 15.18b). A third, inner, membrane is used to create ducts along which to circulate hot or cold air. The terraces of the buildings can be used as floors, since they are under cover, thus increasing the usable areas of the buildings. By the same token, "open air" cafes and other rest areas can be accommodated under the membrane roof and trees and grass can grow there, as proved by the staff

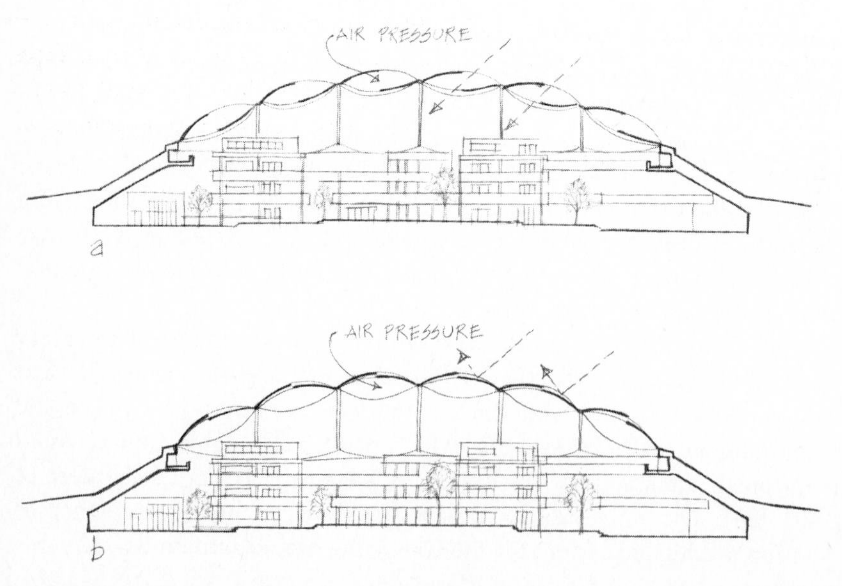

15.18 TRIPLE MEMBRANE THERMAL BALLOON ROOF

of the New York Botanical Gardens. The nine-acre village built under the roof can provide housing, offices, and services for 12,000 occupants with a saving of 17 million dollars in its 30-year life-cycle cost of 89 million dollars. All the strictest building code requirements are met by the building, which in addition has an interesting safety feature, actually tested by a storm in a small pavilion erected in Boston on the occasion of the Bicentennial Celebrations. While the Boston storm destroyed a balloon roof over the M.I.T. sports field, the cylindrical balloon for the Prudential Bicentennial Pavilion also collapsed, but was designed to act as a tent in the collapsed configuration and did not suffer major damage. Similarly, in the Federal Building proposal heavy cables, parallel to the 1,000-foot dimension of the cylindrical roof, are strung above the terraces of the covered buildings and, if at any time the roof were to collapse, the roof cables would be supported by the longitudinal cables and the roof would become a tent. The project, labelled MEG 2 for Megastructure 2, is practically feasible and has introduced a new dimension in architecture, that of landscaping an interior space for an entire town.

Can the dream (or is it the nightmare?) of a covered town be realized? There are no technical obstacles to prevent it. A proposed City in the Arctic for up to 45,000 inhabitants, designed by Frei Otto and Ewald Bubner, already has under consideration a totally climate-controlled membrane over seventy-one acres. It is the prayer of most of us that the surface of the earth be maintained pleasantly habitable, but if population centers are to be built in extreme climates or if man were to be so mad as to endanger life on his planet, technology would nonetheless be ready to shelter him. It is to be hoped that the development of pneumatic structures prompted by such need will remain limited.

16 The Hanging Sky

Domes and Dishes

Mankind has built domed structures for over two thousand years and learned to attribute to their geometry very particular psychological properties. As a source of messages the architectural dome is related to the celestial sphere, which can be interpreted as a dome of infinite magnitude covering the whole of humanity. The feeling of protection given by the sky is more obvious at night when the stars seem either to hang from a spherical ceiling or to be holes pierced in it. This feeling lingers in daytime, when the spherical quality of the sky is less obvious and yet made present by the apparent circular path of the sun. No other surface can give the same feeling of protection because the sphere is the only surface that comes down equally all around us. Moreover, a human being under a large dome-shaped roof has the feeling of being at its center, of being the all-important point in the space covered by the man-made sky. The dome sends a complex and ambiguous message, composed of amazement, awe, and serenity.

If the dome as we normally think of it is 2,000 years old, only recently has mankind perfected its opposite—the upside-down dome. Such a dish roof must be built of steel, or another material capable of withstanding tension, and could only be supported in large dimensions by means of steel cables, the strongest structural material devised so far. Such cables are made by assembling a very large number of steel wires until a sizable, immensely strong cable of circular shape is obtained. As we have seen in

Chapter 10, this American development of the late nineteenth century has been essential to suspension bridges, the longest bridges in the world, and has been recently adapted to this newest kind of large roof, the suspended dish.

What feelings are elicited by these immense surfaces with a curvature opposite that of the dome? A circular dish roof with the lowest point at its center curves evenly upward in all directions. While a dome limits the visual freedom of its occupant by coming down all around its boundary, the dish, seen from the center of its underside, has no visual boundary and seems to rise infinitely, reaching for the sky. An open-sided space covered by a dish roof draws the outside into it and concentrates all at its center (Fig. 16.1).

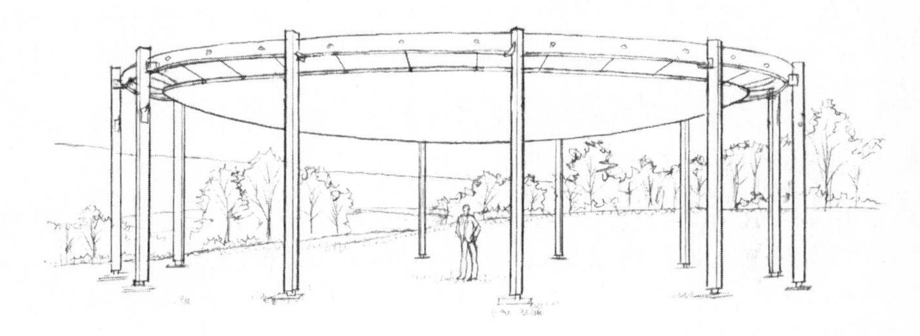

16.1 AN OPEN SIDED DISH ROOF AT CAMP COLUMBIA, LITCHFIELD, CONN.

The message of a dish, however, is different when it spans a closed-in space (Fig. 16.2). Visitors do not occupy its center; we sit at its periphery, and our line of sight is channeled toward the center of the area. Reciprocally, whatever happens there reaches us centrifugally at the perimeter. If the periphery of the area is kept dark and light is concentrated on its center, a mesmerizing centripetal effect is produced.

The space thus defined by an upside-down dome is ideally suited to the mass spectacles so popular among all the cultures of our century. What Dionysus was to the Greeks, the performer is for today's masses, clad in magical light at the center of the space, the dish roof drawing down to him the attention of thousands. Once again, as in all great architecture, light and form are ideally combined for the purpose at hand.

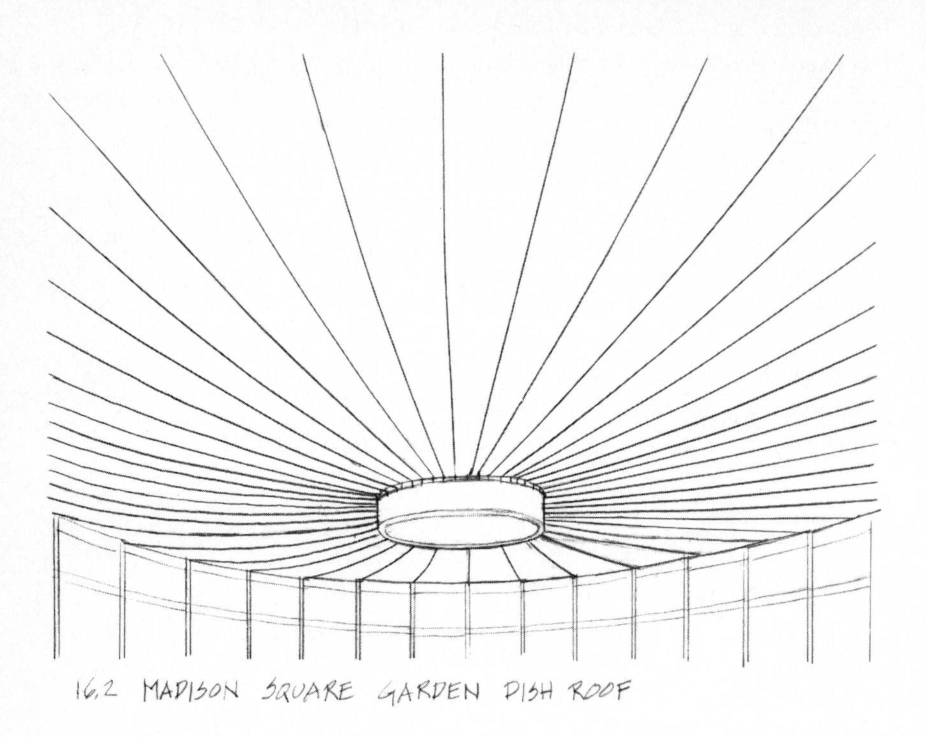

16.2 MADISON SQUARE GARDEN DISH ROOF

We now have dish roofs with diameters of over 400 feet covering three or more acres of seats and over 20,000 spectators. Their technology would easily allow a doubling or tripling of this number. How are they built?

The Hanging Dish

Their essential components (Fig. 16.3) are a large outer compression ring, a smaller inner tension ring, and a number of radial cables connecting the tension ring to the compression ring. The outer ring is supported on columns or, at times, on a circular wall; the inner ring hangs from the radial cables at the center of the space, always at a level lower than that of the outer ring. Segmented slabs (usually of reinforced concrete) placed athwart the cables constitute the surface of the dish. Since the cables (tensed by the weight of the slabs) pull outward on the inner ring, that ring tends to increase in diameter and its fibers elongate. Elongation is always due to tension, the state of stress that pulls the particles of a material apart. Therefore the inner ring must be built of steel, a material that withstands high tensile stresses. The outer ring, conversely, is pulled in by the cables and tends to shrink. A reduction in length is always due to compression. The outer ring is compressed and can be built either of steel, a material that takes tension and compression equally well, or of concrete, a less expensive material that takes compression but not tension.

The judicious choice of materials and the tremendous strength of steel cables (no less than 1,400,000 pounds would be required to pull apart a three-inch diameter cable) make it possible to span hundreds of feet with a suspended dish roof. But the singular advantage of these roofs is the economy of their construction. While the construction of a dome usually requires the erection of a centering of great expense and complexity, the dish roof is erected without a scaffold. The slabs forming its surface can be fabricated at ground level, while the compression and tension rings are built, and then hoisted into place on cables previously strung between the rings. The concrete used for the slabs is only a few inches thick, since the slabs span small distances between the radial cables. Moreover, spacing the cables closely allows the slabs to be of small thickness *whatever* the span of the roof. (The thickness of a dome, on the other hand, must increase with its span.)

Dish roofs are so light that a designer would ordinarily have to worry about the danger of potential vibrations under the action of wind gusts, but a very ingenious method has been devised to prevent these vibrations without increasing the weight of the roof. We owe to the late Uruguayan engineer Leonel Viera this last significant improvement in the cable roof, which he employed in building a stadium in Montevideo,

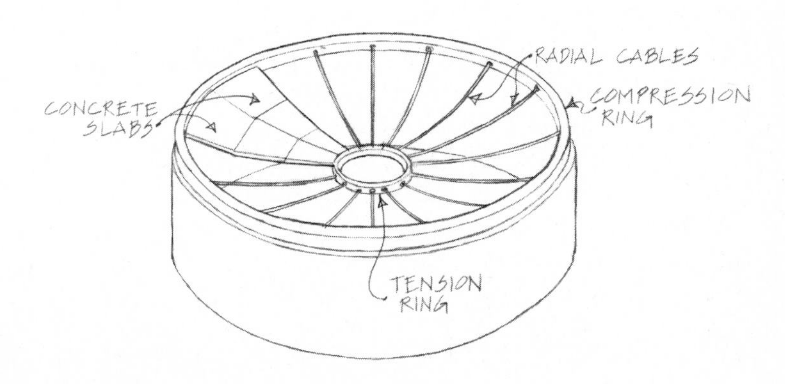

16.3 DISH ROOF STRUCTURE

Uruguay, in 1957. His method aims, like all others, at stiffening the cables and permits an economy and simplicity of construction never reached before. Like most brilliant ideas, the principle of the Viera roof is so elementary that one wonders why it was not thought of years before, particularly since all the basic concepts and technologies required for its realization had been available for a long time.

As usual, Viera set on the cables a concentric series of concrete slabs of varying trapezoidal shape, more and more thin-wedged as they approached the smaller inner ring. But then, rather than grouting and glueing the joints between the slabs at this time, as everybody else had done before him, Viera loaded each slab with an additional ballast of bricks, thus increasing the weight on the cables and stretching them further. Only then did he grout, with a good cement mortar, the radial and circumferential joints between the slabs (Fig. 16.4). When the grout had hardened, making the slabs into a monolithic dish of concrete with the stretched cables embedded in it, Viera removed the ballast of bricks. At this point, the roof tended to move up under the reduced load but could not do so since the cables, grabbed by the solid grout, were prevented from shortening. The roof became a monolithic, prestressed dish of concrete, much more rigid than if it had been tensed by the slab weights only, and built without costly supporting formwork. (Even the labor-consuming operations of ballast-loading and unloading have been done away with lately by using in the grout special "expansive" cements that increase in volume while setting, thereby post-tensioning the cables by their own expansive action.) Viera's stadium (Fig. 16.5) is 310 feet in

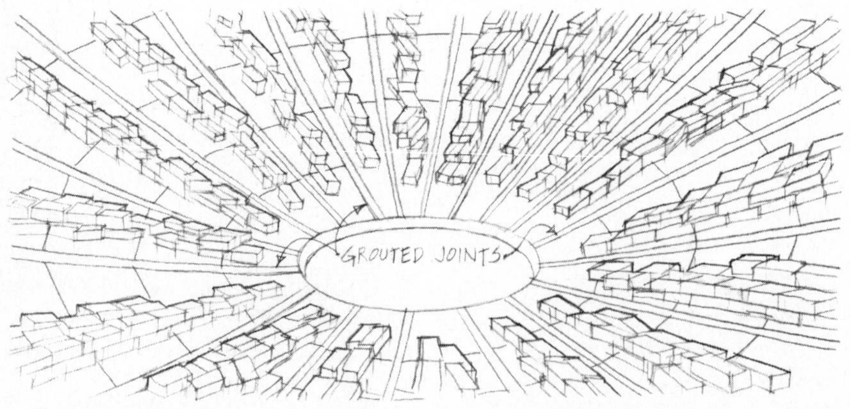

16.4 GROUTING THE SLABS OF A BALLASTED VIERA ROOF

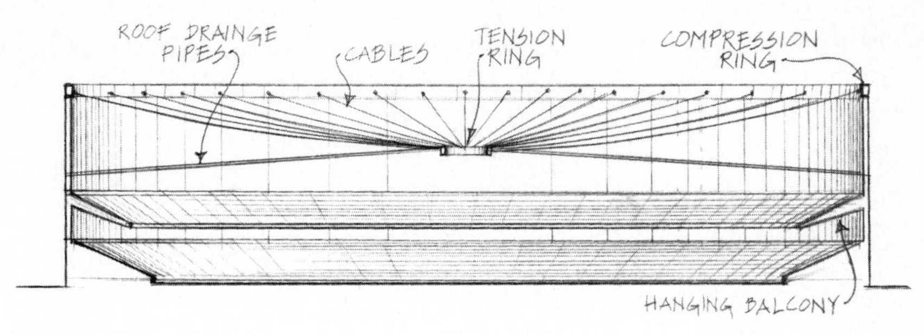

16.5 THE VIERA ROOF IN MONTEVIDEO, URUGUAY

diameter, has slabs two inches thick, an outer concrete ring compressed by the pull of the cables, and an inner tension ring of steel eighteen feet in diameter. The outer wall of the stadium supporting the concrete ring is sixty feet high and only eight inches thick, and the inner steel ring, initially supported on a light scaffold, hangs from the cables of the completed roof. The eight-inch outer wall even supports in addition two sets of circular balconies!

Viera's structural masterpiece encountered an almost unsurmountable objection when first proposed. How does one dispose of the rain water accumulating in the dish, the weight of which may be much larger than the weight of the roof itself? Viera let the water flow down and out of the stadium through four inclined pipes hanging from the inner steel ring. But this was found unacceptable from an aesthetic point of view by many architects. It was then proposed to pump the water over the rim of the roof, as soon as it starts accumulating at the level of the lower tension ring. "But," critics asked, "what if the pump fails?" The obvious answer—a back-up, fail-safe pump—was then countered by the new question, "What if the electricity fails?" Again obviously, one would have gasoline generators ready to activate the pumps. At long last the objections stopped and the Viera system was adopted, in a variety of forms, all over the world. These persistent objections encountered by Viera show the conservative bent of members of the construction guilds and the extreme caution required for the introduction of new ideas in the field of structures. We have seen similar difficulties in the acceptance of pneumatic membrane structures.

Another solution to the water-disposal problem was adopted in two outstanding applications of the dish roof principle in the United States—Madiscn Square Garden in New York (Fig. 16.6) and the Forum Sports

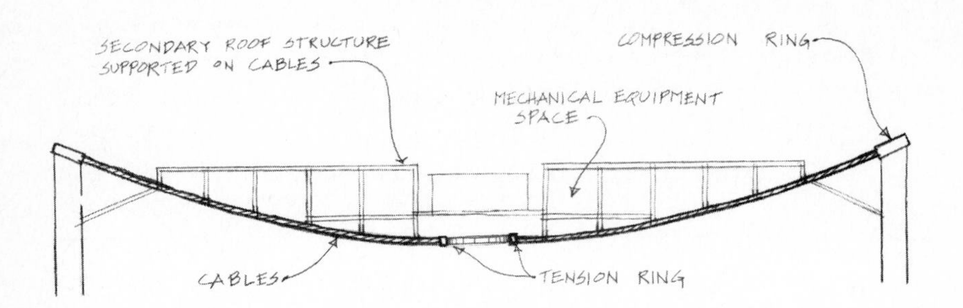

16.6 SECTION OF MADISON SQUARE GARDEN, IN NEW YORK CITY

Arena in Inglewood, California (both designed by Charles Luckman Associates). In both cases the cables support, above the dish surface, a one-story circular building housing the air-conditioning and other mechanical systems as well as the central lighting system for the building. The roof of this additional one-story building is curved downward so that the water runs radially to its perimeter, where it is channelled to the boundary of the building.

While the circular shape is the most logical for a hanging dish, a variety of other shapes have been used that increase the range of expression of these roofs. Elliptical dishes have been built without a center ring by using two nets of cables at an angle to each other (Fig. 16.7). Pure compression and tension rings can also be used with an elliptical dish, if the inner ring is set at one of the foci* of the ellipse. While the circle is a perfectly centered shape, ellipses can be built with any ratio of long to short span. The greater this ratio, the more ecentric the location of the foci. The eccentricity of the inner ring gives a totally different shape to the dish and focuses the sight lines on a point off the center of the covered area. A new dimension, which can be exploited for a variety of functional purposes, is thus added to the classical hanging dish: in spite of the noncircularity of the roof, the outer ring is purely compressed and the inner ring purely tensed, developing the same structural advantages and material economies of a circular roof. Elliptical domes were built in the eighteenth century to cover the crossings of some Baroque churches,

* The foci of an ellipse are two points on its longer axis, having the property that the distances of any point on the ellipse from the foci add up to the same length.

but only in our time have elliptical dishes become a reality in widening the architectural potentialities of hanging roofs.

An ingenious solution to the rainwater problem, suggested and used by Lev Zetlin, has led to a structural variation of the dish roof (Fig. 16.8). This has not one but two inner tension rings, interconnected and set one above the other to form a hub. The lower ring sits, as usual, below the level of the outer compression ring; the upper is above it. Cables are strung in two sets from the compression ring, one connected to the upper tension ring, the other connected to the lower tension ring. Spreader struts are set vertically between the two sets of cables; by spreading them apart, all the cables are put in tension. The resulting structure is nothing but a *horizontal bicycle wheel* with the rim represented by the outer compression ring and the hub by the two interconnected tension rings. The surface defined by the lower set of cables would be a dish, but since panels are not supported on them, such a dish surface does not materialize. Panels are supported, instead, on the upper set of cables defining a dome surface on which the rainwater flows outward to the perimeter of the building.

A totally different cable surface can be obtained if a single inner tension ring is set higher than the compression ring. In this case the tension ring must be supported on a center column, since it does not hang from

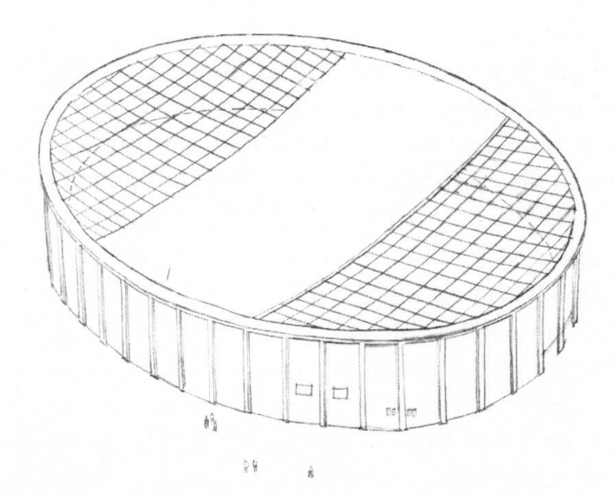

16.7 SUSPENDED ELLIPTICAL ROOF WITH RECTANGULAR CABLE NET

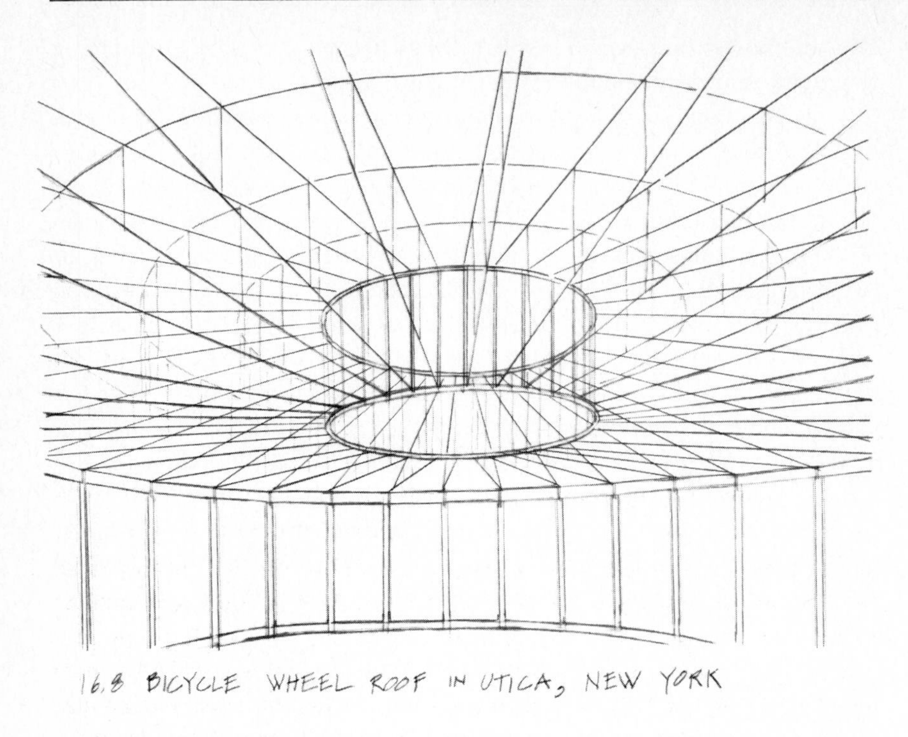

16.8 BICYCLE WHEEL ROOF IN UTICA, NEW YORK

the cables. Although the construction procedure used to erect such a roof is essentially the same as that used for the dish roof, the resulting shape is a circular tent supported on a center pole (Fig. 16.9). Obviously such a shape cannot be used for an auditorium or a sports arena, as the center column breaks up the interior space. On the other hand, if the cables are almost horizontal when they reach the outer compression ring, their pull on it will be practically horizontal, and the entire vertical weight of the roof will be carried by the center column. Thus the outer ring can be supported on widely spaced columns and the periphery of the building can be open to the outside. This is an essential requirement for buildings like hangars or bus depots, and roofs of this type have been built in the USSR with diameters of up to 500 feet.

Finally, it may be mentioned that in a dish or a tent roof the structural function of the cables and the enclosing function of the slabs can be combined by building the slabs out of thin steel sheet and welding them together. This steel "membrane" becomes both cables *and* slabs and presents the utmost integration of architectural and structural needs.

One may wonder what the ingenuity of humankind will invent to satisfy the future needs of our culture. Much as we praise individualism

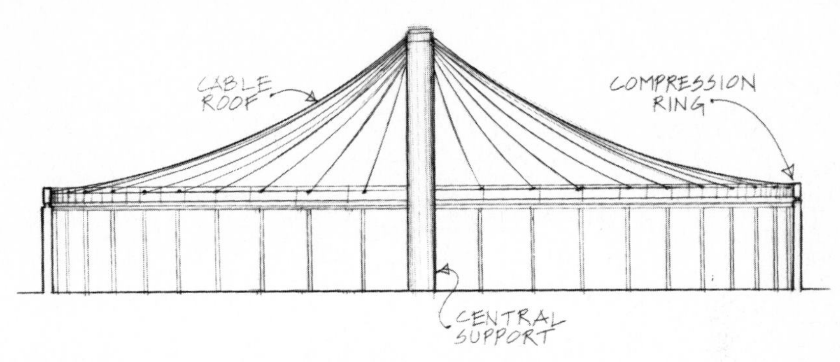

16.9 CABLE ROOF ON CENTER SUPPORT

and independence, the urge to gather in large numbers and thus increase the emotional enjoyment of many experiences is so basic that it appears to be a natural human trait. The Coliseum, with its hanging tent roof protecting the spectators from the harsh Roman sun, allowed people to enjoy spectacles which today we consider inhuman. Madison Square Garden allows 20,000 people to enjoy the somewhat less inhuman show of a man trying to beat another into insensibility or the edifying show of people gathering for spiritual exaltation beyond words. One may dream of the development of new steels of immense strength and of new plastics as strong as steel and with optical properties varying with the light impinging on them. And one may dream of dish roofs, made possible by such technological developments, under which hundreds of thousands of people gather for the enhancement of their deepest experiences.

Perhaps the message of such man-made skies will be—as it was at Hagia Sophia—simply the brotherhood of man.

17 The Message of Structure

Semiotic Messages

There can be structure without architecture, as in any machine, but no architecture without structure. There can be aesthetics without architecture, as in any painting, but no architecture without aesthetics. But is there an influence of structure on aesthetics? Should we be interested in the aesthetics of structure? To answer this question, we may ignore the definitions of "the beautiful" given by aestheticians through centuries and simply note that aesthetic tenets have changed through the ages. A piece of architecture once considered a masterpiece is often demoted later and labelled second-rate, or it may go in the other direction, from oblivion to fame. Aesthetic tenets are dynamic; nothing is absolute about them. And yet, mankind has always tried to achieve aesthetic results, even in its most humble artifacts, because the satisfaction of aesthetic feelings is one of the basic needs of humanity. On this basis alone, we cannot ignore those aspects of structure that are influenced by and, in turn, influence the beauty of a building.

It is easy to prove that aesthetically satisfying buildings can be designed even if structural laws are totally or partially ignored. For example, the conversion of a wooden temple into a stone structure led the Greeks to the creation of one of the masterpieces of architecture, the Parthenon (Fig. 17.1), though judged from a purely structural viewpoint the Par-

17.1 THE PARTHENON

thenon is anything but "correct." * On the other hand, some engineers have preached that one can ignore aesthetics because, if a building is correctly designed structurally, beauty will shine from the correctness of the structure. The innumerable examples of "correct" structures which many consider ugly (Fig. 17.2) disprove this theory. Engineers like Nervi or Maillart have designed aesthetically wonderful structures because their innate feeling for beauty guided them even beyond their structural genius (Fig. 17.3).

Moreover, in considering the aesthetics of a building, one must carefully distinguish between those buildings in which the structure is relatively unimportant, and those in which the structure is essential to the appearance of the building. While a one-family house can be built in many shapes, of wood, steel, bricks, or concrete, the shape and materials of a suspension bridge are almost uniquely determined by structural requirements. All buildings between these two extremes have a structural component that is bound to influence, in varying degrees, their appearance

* Because stone cannot span as great a distance as wood, the diameter of the columns had to be grossly increased to reduce the span of the new stone beams.

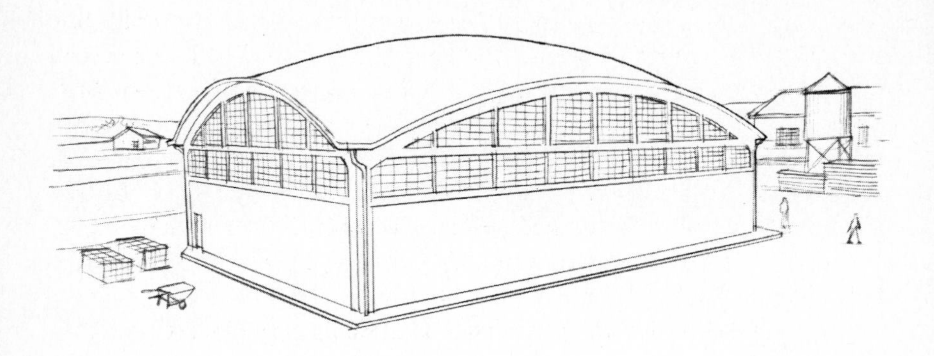

17.2 A "CORRECT" STRUCTURE

and thus influence the aesthetic response of their users. By the same token, constraints due to aesthetics, imposed by the designing architect or by fashion, often influence the structural solution adopted by the engineer, and it is through the interplay of these two personalities of different training that a final compromise solution is achieved.

During the last few years in the discussion of aesthetics a new point of view has become widely accepted by aestheticians, architects, engineers, and other professionals involved in structures—a building has a *semiotic* message. A branch of philosophy developed over the last seventy or eighty years, semiotics considers any and all products of human activity from the viewpoint of communication and particularly of nonverbal communication.

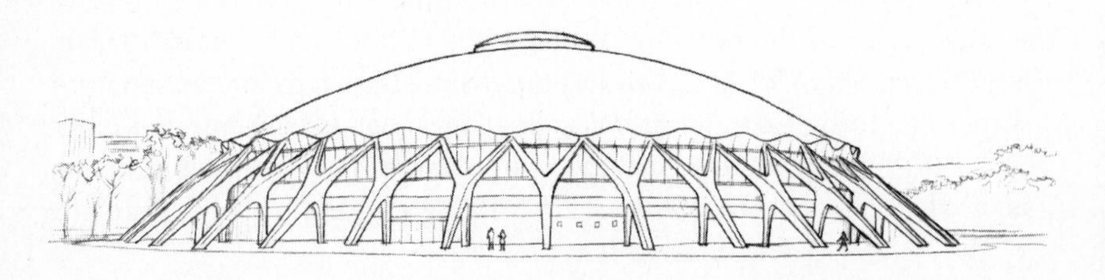

17.3 NERVI'S PALAZZETTO DELLO SPORT IN ROME

Communication takes place by means of messages. Verbal communication is usually concerned with the *meaning* of the message and uses a language to achieve its purpose. Thus, a verbal message is totally dependent on the culture in which the language is understood. Nonverbal communication can also be directly related to a specific message, one that could as well have been expressed verbally. A road sign, which is internationally understood, stands for a specific message intentionally stated—for example, "do not park." But a nonverbal message may also be the by-product of an artifact whose main purpose was *not* to send out a message, but to satisfy a given function. Clothing, for example, is made to cover the human body, but is also used to signify status.

In architecture we often find both kinds of nonverbal messages. The pushbuttons in an elevator or the windows in a building perform specific functions, while at the same time sending out messages. The messages of the elevator buttons have to do exclusively with their function of moving the elevator to a specific level. The window, on the other hand, through its shape and dimensions, may indicate something other than its intrinsic purpose of transmitting light. The barred windows of a jail speak clearly, and the ornamented windows of a Renaissance palace state unequivocally the status of the personage occupying the palace (Fig. 17.4).

17.4 FARNESE PALACE IN ROME

It is also obvious from these elementary examples that, like its verbal counterpart, the semiotic message is always deeply embedded in a culture, A savage entering the lobby of a modern high-rise building would not understand the message of the elevator buttons; neither would a citizen of the jungles of South America understand the social status expressed by the windows of a Renaissance palace. Moreover, like any verbal message, the nonverbal message varies in meaning with time. The message of the pyramids was basically religious at the time of the Pharaohs. It became magic and scientific for later explorers, military and glorious for Napoleon and his troops, and is now a mixture of artistic, sociological, and structural messages for modern visitors.

The use of structure in architecture can introduce a semiotic message in two ways. A building may depend essentially on its structure to achieve aesthetic results, though this may not be apparent to the layman when the structure is hidden. On the other hand, in those buildings where the structure is almost uniquely determined by laws of nature, the message is *strictly related to structural action* and has a meaning of its *own.*

It is common belief that purely structural messages originate in our *intuitive* understanding of structural behavior, which stems both from our

17.5 THE STRUCTURE OF A TREE

17.6a CRETAN COLUMN　　17.6b DORIC COLUMN

daily physical experience with structural actions and from our perception of structural forms in nature. Consider raising water from a well. This activity, which goes back to the beginning of civilization, gives us an immediate understanding of the tensile properties of natural or artificial ropes. Similarly, the breaking of a tree branch against our knee or the sensation in our muscles when we go up or down a slope make it clear that physical observations as well as muscular reactions gave mankind a feeling for structural action much before a quantitative grasp of it was obtained through science.

The visual perception of natural structures has been basic in extending our intuitive understanding of man-made structures. The branches of a tree (Fig. 17.5), acted upon by their own weight and snow or wind, suggest the shape and behavior of cantilevers, with their larger dimensions at the root and their slender dimensions at the tip. The shape of a tree trunk introduces us to the requirements of gravity loads in tall buildings, accumulating from top to bottom. Because of these primordial experiences we feel an instinctive puzzlement at the sight of Cretan columns (Fig. 17.6a) larger at the top than at the bottom, but we accept naturally the shape of a Greek Doric column (Fig. 17.6b). Similarly, as

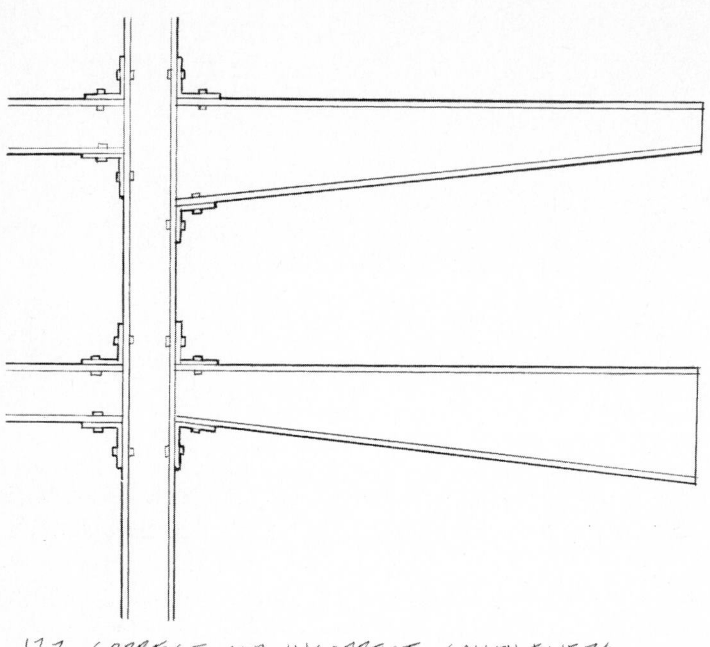

17.7 CORRECT AND INCORRECT CANTILEVERS

already mentioned in Chapter 1, we would consider a cantilever beam, with a larger section at its tip than at its root, "ugly" (Fig. 17.7), since it is an insult to the kind of structural behavior, of which we might not be conscious, but to which nature has exposed us.

Similar reactions of shock occur whenever we are confronted with, say, a large mass in the shape of an inverted pyramid. Mountains, due to the action of gravity, are shaped as right-side-up pyramids. The Egyptian pyramids have a shape, geometrically idealized, but basically identical to that of all the mountains we have ever seen. But a modern building in the shape of an inverted pyramid (Fig. 17.8) does not "say" to the layman how and why it stands up: it tells him that some "trick" has been used to achieve an "unnatural" result. This unnaturalness elicits in us a sense of uneasy surprise rather than the feeling of balance related to "honest" structural behavior.

Natural arches have taught us that when stone is used to span a gap, a downward curvature is needed to achieve the goal. Stone is strong enough in compression to support an entire mountain, and downward curvature creates compression. Three-dimensional structures like domes can be understood in a similar manner. We need only refer to natural

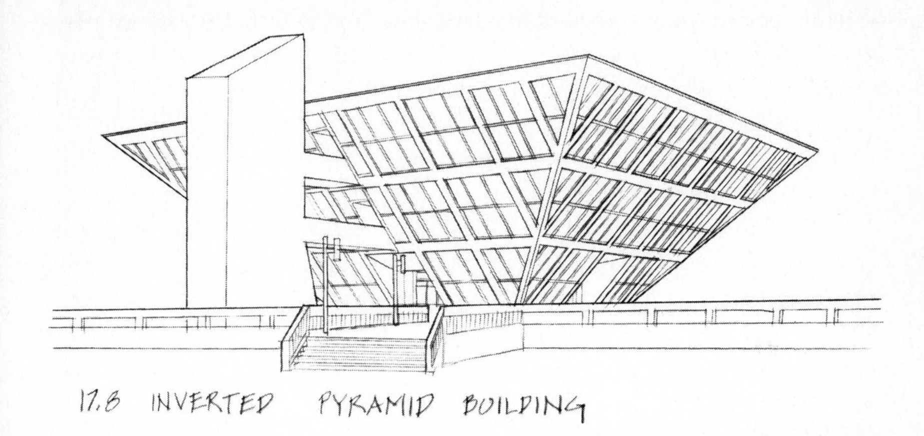

17.8 INVERTED PYRAMID BUILDING

caves, where curved inner surfaces give us a feeling for arch action in space. Sea shells (Fig. 17.9) are a symbol of protection, but they also have a strong aesthetic content, whenever they are ribbed in any of the great variety of ways in which nature has shown us how to stiffen a curved three-dimensional surface. The vine catenaries linking tree to tree show us the need for an upward curvature in a tension structure like a suspension bridge, a requirement that modern engineering well understands.

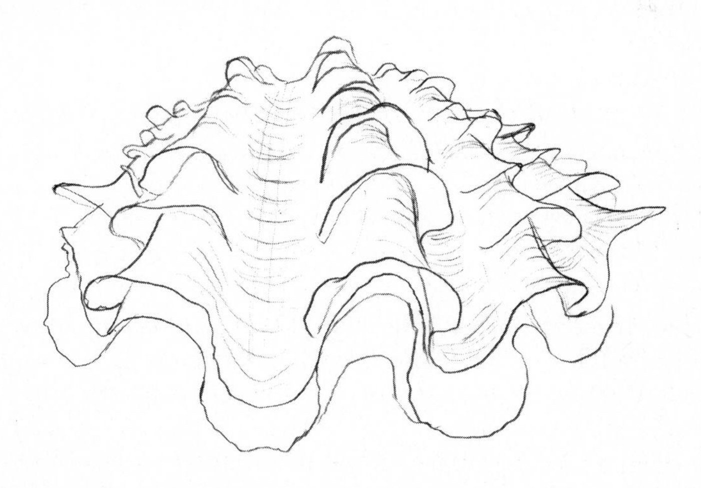

17.9 RIBBED SEA SHELL

It would seem, therefore, that to the layman the semiotic message of structure comes from a series of atavistic intuitions and that the accumulation of these intuitions results in a set of aesthetic responses. This is why the layman considers a correctly designed cantilever beam as "lovely" and a cantilever with incorrect structural dimensions as "ugly" (see Fig. 17.7).

Shifting our attention to some structures which have only recently come into common use—for example, steel frames hinged at their base (Fig. 17.10)—we notice that their structural message is equivocal to the lay reader. Consequently it appears that such structures do not have, as yet, much aesthetic content for the layman nor do they bespeak beauty to the experts. While the curved shape of an arch has a strong aesthetic charge, a frame appears to be a utilitarian structure, and one would be hard put to mention a building which has enhanced its aesthetics by the use of such a structural element.

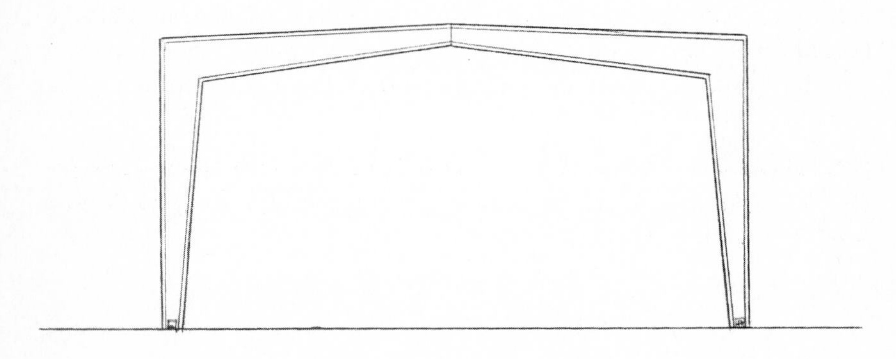

17.10 HINGED STEEL FRAME

A similar reaction is elicited by those *correct* structural shapes that have their justification in subtle physical phenomena. For example, a straight rod with a shape dictated by resistance to buckling (Fig. 17.11) is hardly viewed as beautiful or ugly; most of the time it is looked upon as a machine component. In fact, machines are often considered ugly because while their shapes are totally correct from a structural and a functional point of view, they are so "new" that we cannot grasp them yet as part of the universe of aesthetics. It is true that some artists at the beginning of this century (Fernand Leger, for one) introduced machine-

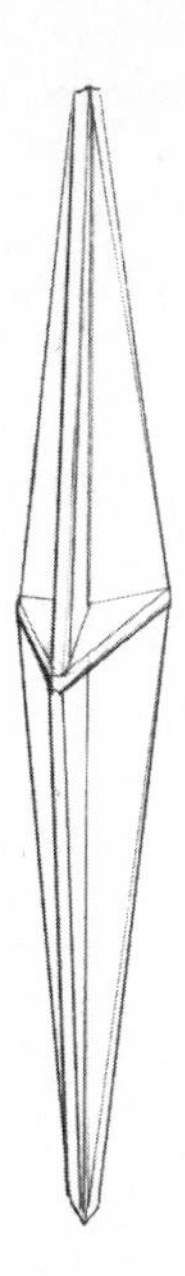

17.11 CORRECTLY SHAPED COMPRESSION ROD

like elements into their paintings. But they were careful to dissociate the painted image from the world of mechanical reality; Picabia's famous machines do not work.

Semiotic Message and Scale

It is interesting to notice that *scale* does not seem to impair the semiotic message of large man-made structures. The reference of the message to the common experiences of the race seems to be unaffected by size, as shown by the comparison of almost any tensile structure to a spider web. Because of the efficiency of tension, tensile structures are always exceptionally light and, whatever their shape and size, are considered "elegant" by the layman (see Fig. 15.3).

It is interesting to notice at this point that no domed structure is lighter than a modern balloon roof (Fig. 17.12). But its natural shape, which should look "elegant" to the inexperienced eye because of its lightness and translucency, is not yet comprehended aesthetically and is more a source of puzzlement than of appreciation. From the outside most

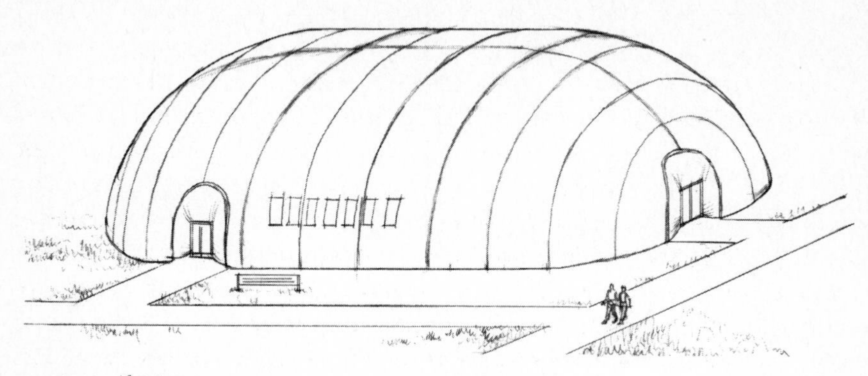

17.12 PNEUMATIC TENNIS COURT COVER

balloon structures spell heaviness, and this is an obstacle to their general acceptance. The traditional message of the stone dome in compression does not allow us, as yet, to understand the recent pneumatic structural message, which therefore confuses us both structurally and aesthetically.

This confusion does not arise with tents. These "pure" tensile structures have their counterparts in nature. They have also been seen by the race for thousands of years. Hence both real tents (Fig. 17.13) and similar tensile roofs elicit the kind of aesthetic appreciation showered on most curved structures modeled after nature.

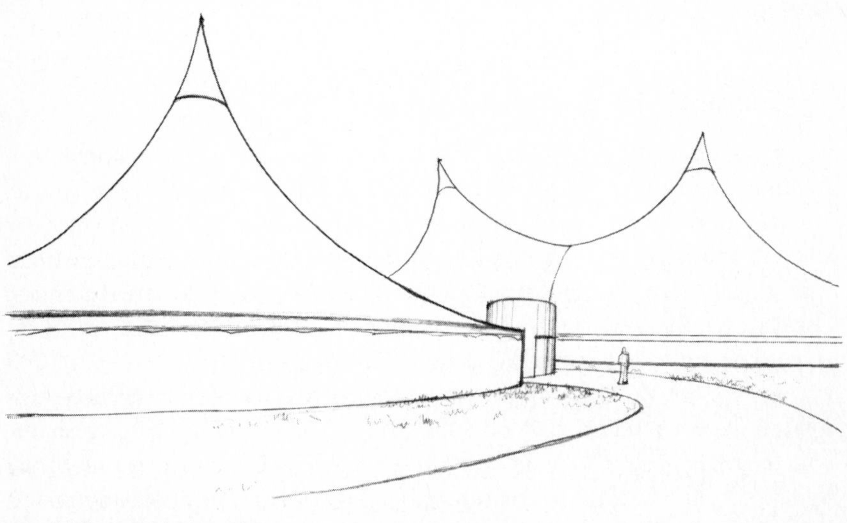

17.13 TENSILE ROOF

One should not consider the understanding of structural behavior a necessary condition for the aesthetic appreciation of a structure. A striking example of the unimportance of structural understanding is given by the shape of a roof universally admired aesthetically while seldom understood structurally. The hyperbolic paraboloid, one of the most efficient structural roof forms, is blessed by a feature often appearing both in nature and in modern sculpture: a saddle shape. From a structural point of view hyperbolic paraboloids can be correctly used in a more or less horizontal position to create roofs, or in a structurally unjustifiable vertical setting like walls (see Fig. 11.30). But their aesthetic message is always one of beauty. The reaction of a twelve-year-old, who had never before seen a roof shaped like the hypar in Figure 17.14, may indicate some of the aesthetic associations dictated by the semiotic message of the hyperbolic paraboloid. He first saw the similarity to a horse saddle, but then on second thought added, "It also looks like a bird flying."

17.14 HYPAR BUTTERFLY ROOF

In some cases even an unconscious understanding of the correctness of a structure may enhance its aesthetic content. In a slab with ribs curved along lines where the plate does not carry loads by twisting (see Fig. 11.7), the pattern of the ribs becomes a source of aesthetic satisfaction even for those who would not dream that twisting takes place in a plate (see Chapter 11).

One wonders if the puzzling message of certain structures will ever permit aesthetic acceptance of their behavior. The shape of prestressed concrete elements, governed by the tension in their invisible tendons,

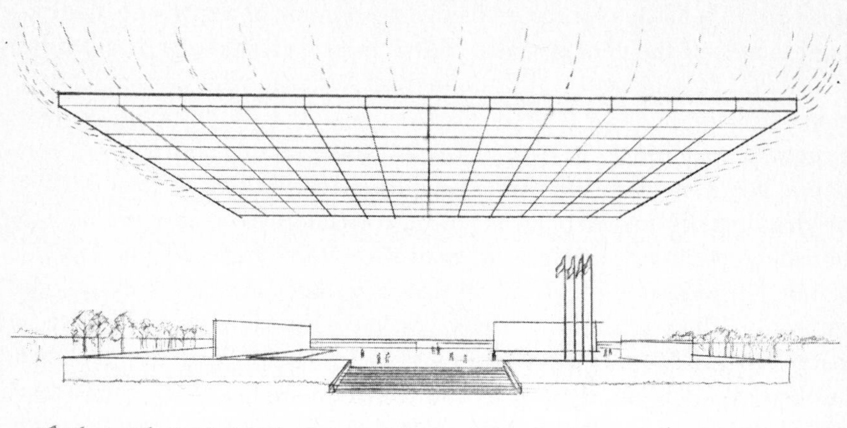

17.15 ROOF SUPPORTED BY ELECTROMAGNETIC FIELDS OF FORCE

contradicts our intuition of a "logical" shape. What will be the aesthetic reaction of future generations to roofs floated in space by electromagnetic fields of force (Fig. 17.15)? These invisible fields have no reference in nature and seem to defy its basic laws. This defiance is both puzzling and shocking since there can be no doubt about the satisfaction from a semiotic message aesthetically *and* structurally understood.

It is this satisfaction, this harmony between the visual need for beauty and the respect for basic laws of nature, that has dictated in the past and dictates even more in the present, the tendency to exhibit the structure of a building. The same satisfaction is the source of our admiration for the Roman vaults, on the one hand, and the John Hancock Insurance Company Building in Chicago, on the other (see Fig. 7.9).

The Varying Semiotic Message

Although the semiotic message of a structure is strictly related to our personal experience and culturally to the experience of the race, within a short number of years our aesthetic appreciation of a given structure can change, as proven by the classic example of the Eiffel Tower in Paris (see Fig. 8.1). As we have seen, this extraordinary steel structure, erected by an engineer of genius, originally had a limited purpose and was to be dismantled at the closing of an exhibition. The campaign against its erection involved some of the most respected representatives of French culture of the time, but it did not take long for the Tower to become not just *one* of the symbols of that center of world culture called Paris but *its very* symbol. And a few years later the Tower, on its own structural steam, became the semiotic symbol of France. In this extraordinary case, the total semiotic message stems directly and uniquely from a structural message. The Eiffel Tower is a structural masterpiece in which almost nothing was conceded to decoration, nothing used to hide its necessary sinews. Its acceptance indicates not only an amazing reversal of public opinion but the possibility of a pure aesthetic message emanating from a pure structure. A similar interaction between structural and aesthetic needs preserved the nakedness of the towers of the George Washington Bridge (see Fig. 10.6) despite the opposition of a large part of the New York intelligentsia and of the designing engineer himself—another indication of the rapid change in aesthetic appreciation of the semiotic message of a structure.

The Beaubourg Museum has just recently been inaugurated in Paris (Fig. 17.16). The dismay of the world at the erection of an art shelter with an aesthetic message based not only on its visible structure but on its mechanical systems must be understood in the light of our recent historical past. One cannot forecast that the Beaubourg will become the new

17.16 THE BEAUBOURG MUSEUM IN PARIS

symbol of Paris or of modern art. But one should not be surprised if the incorporation of mechanical elements into its aesthetic message were to lead to a widening of the vocabulary of architecture and become accepted as a matter of course within a few years. After all, if the last operas of Mozart were criticized as "noise, not music" by Viennese critics of the time, if the early twentieth century French painters were called "the wild beasts," if Joyce's *Ulysses* was bitterly attacked, why should the aesthetic message of structural and mechanical systems enjoy a different reception?

It may be surprising to realize, at the end of this rapid excursion through the field of architectural structures, that such a highly technological field has contributed and will continue to contribute to our innate need for beauty. To those of us who cannot live without beauty, this is an encouraging thought. The separation of technology and art is both unnecessary and incorrect; one is not an enemy of the other. Instead it is essential to understand that technology is often a necessary component of art and that art helps technology to serve man better. Nowhere is this more true than in architecture and structure, a marriage in which science and beauty combine to fulfill some of the most basic physical and spiritual needs of humanity.

Index

Catalog